Desorption Mass Spectrometry
Are SIMS and FAB the Same?

Philip A. Lyon, EDITOR

3M

Developed from a symposium sponsored by
3M,
the National Science Foundation Midwest Center for
Mass Spectrometry at the University of Nebraska—Lincoln,
and
the National Science Foundation Regional Facility for
Surface Analysis at the University of Minnesota,
St. Paul, Minnesota,
October 7–10, 1984

American Chemical Society, Washington, D.C. 1985

52828074

SEPIAE
CHEM

Library of Congress Cataloging in Publication Data

Desorption mass spectrometry.
  (ACS symposium series, ISSN 0097–6156; 291)

  Includes bibliographies and indexes.

  1. Field desorption mass spectrometry—Congresses.
2. Secondary ion mass spectrometry—Congresses.

  I. Lyon, Philip A., 1945–   . II. American
Chemical Society. III. Series.

QD96.M3D47   1985   547.3'0873   85–20151
ISBN 0–8412–0942–1

QD96
M3 D471
1985
CHEM

ACS Symposium Series

M. Joan Comstock, *Series Editor*

Lawrence A. Casper
Specialty Editor in
Inorganic Materials Science

This book was acquired for publication through the efforts of Lawrence A. Casper acting in behalf of the American Chemical Society as a Specialty Editor. Dr. Casper is an employee of Honeywell at the Solid State Development Center in Plymouth, Minnesota.

Advisory Board

FOREWORD

The ACS SYMPOSIUM SERIES was founded in 1974 to provide a medium for publishing symposia quickly in book form. The format of the Series parallels that of the continuing ADVANCES IN CHEMISTRY SERIES except that, in order to save time, the papers are not typeset but are reproduced as they are submitted by the authors in camera-ready form. Papers are reviewed under the supervision of the Editors with the assistance of the Series Advisory Board and are selected to maintain the integrity of the symposia; however, verbatim reproductions of previously published papers are not accepted. Both reviews and reports of research are acceptable, because symposia may embrace both types of presentation.

CONTENTS

PREFACE

Mass spectrometric instrumentation and the capabilities for analysis of organic and organometallic molecules have undergone revolutionary advances in the last 3 years. Perhaps the most notable advances are in the area of volatilization and ionization of samples. In 1981, a new ion source was developed for a conventional high-resolution magnetic mass spectrometer that allowed the chemist for the first time to analyze organic compounds that were ionic, nonvolatile, or thermally unstable. This ion source used a fast atom beam to generate the organic ions. Tremendous growth has occurred in the use of the fast atom bombardment (FAB) source in combination with conventional mass spectrometers, both in academic research and in industrial problem solving.

The technique of FAB mass spectrometry (FABMS) has many similarities to that of secondary ion mass spectrometry (SIMS). The basic designs of the ion sources are similar, and these sources may share a common mode of generating ions. However, many researchers using FABMS consider their work to be original discoveries and disregard a wealth of knowledge in the field of the surface scientist. The SIMS method is significantly ahead of FAB in its development. Those doing FABMS have much to learn from SIMS studies. This condition of two analytical techniques advancing down parallel paths without any interaction slows the progress in both fields and, more importantly, prevents full utilization of these techniques.

The symposium upon which this book is based was held to encourage an open dialogue between researchers in the fields of SIMS and FABMS. The intent of the symposium and this book is to provide the basis for an interdisciplinary discussion of both the theoretical and applied aspects of these surface analytical techniques. The goal is to demythologize the subject of particle bombardment and also to bridge the gap that often exists between researchers in the fields of SIMS and FABMS. Scientists with a modest knowledge of mass spectrometry should gain a clearer understanding of desorption techniques and how they can be applied.

The book is organized into three sections. The first contains the most recent views on fundamental aspects of particle bombardment. Discussions of ^{252}Cf plasma desorption and laser desorption mass spectrometry have been included for comparison. The second section addresses the issues involved in instrument design, covering work on liquid metal and FAB ion guns. The last part presents representative applications of these bombardment methods.

The historical development of particle bombardment was presented by R. E. Honig as a retrospective lecture at the 32nd Annual Conference on Mass Spectrometry and Allied Topics in San Antonio, Texas, in 1984. This excellent lecture has subsequently been published, and I recommend it for those who wish additional background on the topic [Honig, R. E. In *The 32nd Annual Conference on Mass Spectrometry and Allied Topics— Retrospective Lectures;* Finnigan, R., Ed.; American Society for Mass Spectrometry: East Lansing, MI, 1984; Honig, R. E. *Int. J. Mass Spectrom. Ion Phys.* **1985,** *66,* 31–54].

Acknowledgments

Those serving with me on the organizing committee and contributing greatly to the success of the symposium were M. L. Gross, University of Nebraska; R. M. Hexter, University of Minnesota; and J. A. Leys and W. L. Stebbings, 3M.

I extend my sincere thanks to those who handled the many details associated with the symposium. Sharon Hunt handled preregistrations and mailed information to the conferees. Gary Korba organized the poster session and abstracts. On-site registation and other details were covered by Joe Schroepfer, Frank Dehn, Diane Schroepfer, and Deanna Stebbings.

Finally, I gratefully acknowledge the generous support from 3M, Kratos Analytical, Hewlett-Packard, Extranuclear Laboratories, National Science Foundation, Nicolet Instruments, and Perkin–Elmer Physical Electronics Division.

PHILIP A. LYON
Central Research Laboratories
Building 201–BS–05
3M—3M Center
St. Paul, Minnesota 55144–1000

Molecular Secondary Ion Mass Spectrometry

Steven J. Pachuta and R. Graham Cooks

Department of Chemistry, Purdue University, West Lafayette, IN 47907

Progress in molecular secondary ion mass spectrometry
(SIMS) is presented, with emphasis on applications and
the mechanism of ion formation. The mechanism involves
three processes: (1) energy conversion at the surface,
(2) ion/molecule and electron transfer reactions in the
selvedge, and (3) unimolecular dissociations of
internally excited gas phase ions. The role of matrix
effects in mechanistic studies is discussed, as are
experiments which use chemical reactivity to gain
insights into mechanism. The use of tandem mass
spectrometry (MS/MS) in ion structural determinations in
SIMS and other desorption ionization experiments is
illustrated. MS/MS provides evidence for unimolecular
dissociations of gas phase ions, which appear to
underlie much of the fragmentation seen in molecular
SIMS. Cases of strong molecule-surface interactions can
result in dissociation in situ, however, and
examples are collected. Applications of molecular SIMS
in quantitative and trace analysis, chromatography,
studies of ion chemistry, catalysis, and imaging are
reviewed. Developing areas in molecular SIMS include
highly endothermic fragmentations and ion beam induced
surface reactions.

It is a remarkable feature of secondary ion mass spectrometry
(SIMS) that considerable chemical information is accessible through
the procedurally simple physical technique of sputtering.
SIMS--especially under low primary ion flux conditions ("static
SIMS," also known as "molecular SIMS" when applied to
compounds)--provides information on molecular weight and molecular
structure and allows isotopic analysis. The surface sensitivity of
SIMS permits its use in imaging, in monitoring of surface

0097-6156/85/0291-0001$11.75/0
© 1985 American Chemical Society

reactions, and in characterizing the "local atomic structure" of the surface of a complex material. High primary ion fluxes are useful for depth profiling and for analysis of materials dissolved or suspended in liquid matrices. Some characteristics of molecular SIMS are given in Table I.

This review focuses on the phenomenon of molecular SIMS--that is, the physical and chemical bases for its many analytical applications. The applications themselves are also reviewed. The coverage is somewhat historical, emphasizing progress which has come out of this and other laboratories in the past five years. SIMS is discussed in the context of experiments using related desorption ionization (DI) methods, especially laser desorption (LD) and fast atom bombardment (FAB).

It will be helpful to start by considering the status of molecular SIMS as of 1980, with particular reference to a review (1) which summarized progress in molecular SIMS to that date. Perceptions of the basic mechanism in SIMS have changed surprisingly little in the ensuing period, although considerable advances have occurred in experimental and instrumental techniques. In 1980 fast atom bombardment mass spectrometry using liquid matrices did not exist, matrix effects in SIMS were little explored, and there were few SIMS studies of catalysts. Analysis of high molecular weight compounds by SIMS was hampered by the limitations of the quadrupole and low mass range sector analyzers which were used almost exclusively at that time.

While there has been rapid progress in each of these areas, the generation of polyatomic ions is still seen in the terms presented in 1980: (i) the conversion of energy from the form in which it is originally applied into net translational energy of a molecule sufficient to allow it to leave the surface, (ii) the ion/molecule and other chemical reactions which occur at the interface and in the selvedge and which transform the surface molecules of interest into gas phase ions suitable for mass analysis, and (iii) secondary processes which alter the nature of the ion beam after it leaves the interfacial region, especially fragmentation due to unimolecular dissociation of internally

Table I.   Some characteristics of molecular SIMS

Analyzer types	Quadrupole, sector(s), time-of-flight
Vacuum requirements	$10^{-10} - 10^{-6}$ torr
Primary ion	Ar^+, Xe^+, Cs^+, O_2^+ common
Primary ion current	$< 1 \times 10^{-8}$ A cm^{-2} for static SIMS, but higher currents sometimes acceptable
Primary ion energy	500-10000 eV common
Sample composition	Involatile organics, inorganics, and organometallics; semiconductors; adsorbed gases
Physical form of sample	Foils; bulk solids; compressed pellets; frozen and liquid matrices
Sample size	100 μg - 10 ng common; bulk (multi-layer) samples often used
Sputtering yields	1% - 0.01% common
Secondary ion energy distribution	About 3 eV average, depending on sample; no high energy tail as with atomic species
Mass range	Usually below 2000 amu; > 20000 possible
Detection limit	$\leq 10^{-15}$ g for salts
Other features	Surface-sensitive; isotope-specific; can distinguish molecular weight, molecular structure; some capability for depth profiling and imaging

excited gas phase ions. The selvedge (the term is due to Rabalais
(2)) is the plasma formed at and immediately above the surface
during sputtering.

Figure 1 is an early representation of these three regimes
with their distinctive physical and chemical phenomena (3). In
this early picture, energy interconversion was considered as a form
of isomerization--"energy isomerization"--leading to an expression
of the excitation in a form more or less independent of the type of
energy input. Vibrational excitation, especially of the lower
frequency modes corresponding to intermolecular motion, was
considered as the basis for desorption. Activation of surface
phonons expresses these ideas in different currency.

Three types of ionization processes were distinguished (1)
as contributing to the ions observed in molecular SIMS spectra:
direct desorption of precharged materials,
cationization/anionization, and electron ionization (Equations 1-3,
respectively). The equations illustrate overall reactions and do
not attempt to explain detailed mechanistic steps.

$$C^+A^- \ (s) \longrightarrow C^+ \ (g) + A^- \ (g) \qquad\qquad (1)$$

$$M^o \ (s) \longrightarrow M^o \ (g) \xrightarrow{\ C^+\ } [M+C]^+ \ (g) \qquad\qquad (2)$$

$$M^o \ (s) \longrightarrow M^o \ (g) \xrightarrow{\ e^-\ } M^{+\cdot} \ (g) \qquad\qquad (3)$$

Desorption of precharged materials (i.e., salts) is a highly
efficient process, since energy is not channeled into both an
ionization step and a desorption step; previously existing ions are
simply transferred from the solid phase to the gaseous phase. This
effect may be seen in the ease with which SIMS spectra of
quaternary ammonium salts are obtained (4). Derivatization of
zwitterions to yield species with a net charge illustrates the same
point. Cationization or anionization of neutral molecules by
attachment of metal ions, protons, and other charged species is the
second commonly observed ionization process in molecular SIMS
(5). This may involve desorption of neutral molecules

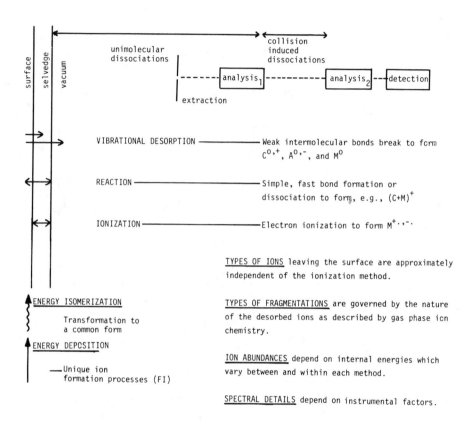

Figure 1.   Early proposal of a unified model for SIMS and other desorption ionization experiments.   Reproduced with permission from Ref. 3.   Copyright 1983, Elsevier Science Publishers B.V.   (First presented at a conference on ion formation from organic solids, Münster, West Germany, 1981.)

concurrently with metal ion production, followed by reaction in the selvedge to form an adduct. Electrons present in the selvedge as a result of secondary electron emission can ionize neutral molecules to give the third type of ionization process seen in molecular SIMS. This process, formation of cation and anion radicals, is not particularly efficient, but it can sometimes lead to abundant ions, for example, in SIMS spectra of polycyclic aromatic hydrocarbons (1).

In addition to those ions formed during or soon after primary ion impact, as in the processes just described, other ions arise through subsequent events. Unimolecular reactions of ions, akin to metastable decompositions in magnetic sector mass spectrometry, occur in the free vacuum. The resulting fragment ions have intensities which contradict the notion that SIMS is a "soft" ionization technique, although some fraction of the ion production events can be so characterized.

In the discussion of mechanism given below, support for the concepts just outlined will be marshalled from newer experimental results which are accomodated by the above model and which call for intact emission of molecules. It should be noted, however, that in the case of a strong molecule–substrate interaction, fragmentation probably occurs directly at the surface. Clearly this is the case when polymers are examined by bombardment techniques. Almost all molecular SIMS experiments employ static conditions in which virgin surface is selected for analysis. Typical maximum ion current densities of 1×10^{-8} A cm^{-2} correspond to ca. 6×10^{-5} ions s^{-1} per surface molecule of 10 $\overset{o}{A}^{2}$ area. This allows sputtering times to exceed one hour before there is a significant probability of examining modified surface material. However, examples are being encountered and will be discussed below where surface chemical reactions ("beam damage") do occur under static SIMS conditions in particular cases. This can produce extensive cleavages of high energy bonds and result in distinctive SIMS fragmentation behavior.

Data will also be given for other desorption ionization experiments which support the general notion of energy isomerization, although no detailed treatment of energy interconversion is attempted. The third aspect of the SIMS mechanism, unimolecular dissociation of isolated gas phase ions, is also the topic of some of the newer experiments reported below, including those which utilize tandem mass spectrometry to characterize these events directly.

Mechanism

The fundamental nature of the desorption process is a continuing subject of controversy. The extensive literature on sputtering of atomic species has led to models which adequately explain most aspects of atomic SIMS (6, 7). Molecular SIMS, however, presents a greater challenge, and the means by which a large biomolecule becomes an ion are less clearly understood. Models have been proposed (8,9), and some current models of molecular desorption are described in the proceedings of this symposium (10-13). Our own qualitative but not untested views are given above and more extensively in the sections which follow. A significant question on which different views have been taken is this: Is fragmentation of stable, strongly internally bonded organic molecules upon primary ion impact significant, or is this overwhelmingly the result of delayed gas phase dissociations of energetic ions? There are dynamical calculations which confirm that such instantaneous dissociations can occur (14), and there are angle-resolved SIMS data which have been interpreted as evidence that they can be a major source of fragment ions (15). Much of the data for bulk samples presented and cited herein, including the similarities in behavior observed when comparisons are made with gas phase processes, lead to the opposite conclusion. Time-resolved experiments in the picosecond range might resolve this issue but are not now accessible.

A feel for the nature of the mechanistic problem can be had by examining the environment near the surface after an impact event using simple calculations and widely-accepted assumptions. It has been noted that a primary ion current density of 1×10^{-8} A cm^{-2}

corresponds to 6×10^{-6} ions $\overset{o}{A}^{-2}$ s^{-1}. This means that, on the average, each area of 6×10^6 $\overset{o}{A}{}^2$ receives one "hit" by a primary particle each second. Because of the very large relative distances and long times between impacts, each must be considered as an isolated event. Consider that a single impact can sputter ten particles, each of mass 200 amu and kinetic energy 2 eV. Using the relation that kinetic energy = $1/2$ mv^2, where m is mass and v is velocity, the velocity of a 2 eV particle of mass 200 amu is 1.39×10^5 cm s^{-1}, or 1.39×10^{13} $\overset{o}{A}$ s^{-1}. The kinetic theory of gases makes possible the calculation of absolute pressure for particles of any kinetic energy through the relation $P = 1/3$ nmv^2, where n is the number of particles per unit volume (16). Since the mass of a 200 amu particle is 3.32×10^{-22} g, the absolute pressure for ten such particles of 2 eV energy is 1.8×10^{10} $\overset{o}{A}{}^3$ torr. If an appropriate volume can be justified, the pressure in this volume can be calculated by simple division. Suppose the selvedge region is a sphere of radius 50 $\overset{o}{A}$; half of the sphere is below the pre-impact level of the surface, and half is above it. The volume of this region is 5.2×10^5 $\overset{o}{A}{}^3$. The pressure in this volume is thus 3.5×10^4 torr as long as the ten particles remain within the sphere--a time of roughly 3.6×10^{-12} s, assuming particles originate at the center of the sphere and there are no collisions.

It should be stressed that these calculations are of the most elementary nature and are meant only to give a feel for what the actual situation may be. If the chosen conditions are varied extensively, however, the basic conclusions remain unchanged. It is interesting to note that even within a volume of 10^{10} $\overset{o}{A}{}^3$ the pressure will still be about 1 torr. In this case particles will have existed for around 10^{-10} s (hundreds of bond vibrations). A recent FAB study (17) found evidence for sputtering from bulk glycerol, based in part on the sputtering of a 10^5 $\overset{o}{A}{}^3$ volume with each primary particle impact (corresponding to ejection of more than 1000 glycerol molecules per Xe primary atom). Very high local pressures are clearly involved in FAB, a situation which supports the assumptions made in our own comparatively conservative calculation. The picture of the selvedge that emerges is that of a very hot, high pressure region in which multiple collisions are possible. This can explain many of the experimental observations discussed below.

Matrix effects. While theoretical approaches to mechanistic studies form a large part of the literature, it is also possible to gain insight through observations based on chemical modification of the system of interest. Modification of the sample matrix provides one chemical approach to mechanistic information. Several different types of matrix effects have been observed. For example, Figure 2 shows SIMS spectra of mixtures of a quaternary ammonium salt and ammonium chloride. The neat compound gives an abundant intact cation at m/z 390, as expected for a precharged species. Rearrangement with loss of cyclohexene to give a fragment at m/z 308 is the dominant fragmentation process observed in the SIMS spectrum of this compound. Similar fragmentations, occurring via elimination of stable neutral molecules, are also observed when quaternary ammonium ions are collisionally activated (18). If NH$_4$Cl is physically mixed with the organic salt at dilutions of 1:2 and then 1:20, an increase occurs in the intact cation abundance relative to the fragment abundance (19). This effect occurs without a decrease in signal-to-noise ratio, even at dilutions as high as 1:1000. At such high dilutions the ratio of m/z 390 to m/z 308 levels off at a value of about 5:1. A tentative explanation for this matrix effect is that the intact cation is initially solvated by NH$_4$Cl. The NH$_4$Cl then forms NH$_3$ and HCl in a desolvation process which serves to take up internal energy from the cation. This loss of internal energy results in decreased fragmentation. The desorption/desolvation sequence suggested seems reasonable in view of observations (see below) of solvent-cation adducts in FAB (20), and of self clustering in FAB spectra of NH$_4$Cl itself (21) and SIMS spectra of low temperature matrices. An alternative explanation is that sputtered cations are collisionally relaxed by interaction with matrix species in the selvedge, a process only subtly different from

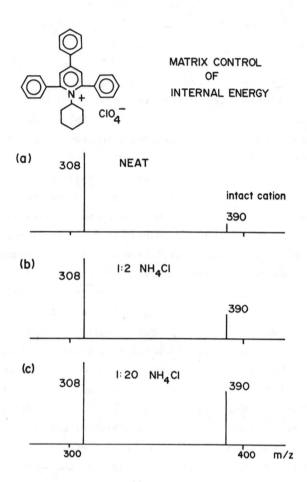

Figure 2.   Effect of dilution in a matrix on fragmentation in
SIMS. Reproduced with permission from Ref. 19.   Copyright
1983, American Chemical Society.

desolvation.   In either event, gas phase dissociations appear to
occur subsequent to intact molecular emission.

A related matrix effect of considerable analytical interest is
the enhancement in absolute ion yield sometimes observed under
conditions of high dilution in a solid matrix (22). Comparison
of the SIMS spectrum of a neat pyrilium salt with that of the same
salt diluted 1000-fold in NH_4Cl shows that the intact cation
signal is observed in about three times greater abundance for the
NH_4Cl-diluted sample.   The threefold increase is observed even
when the absolute amount of salt analyzed in the dilute sample is
one thousand times less than that in the neat sample.   An
additional aspect of this experiment is the persistence of the
enhanced signal.   Ion bombardment yields products for one day in
the NH_4Cl matrix, but for only about one hour in the neat sample
under identical conditions.   Effective desorption of ammonium
chloride, which entrains analyte, is one way of accounting for
these observations.

Independently of the detailed model used to describe these
matrix effects, one must propose that mixing on a molecular level
occurs between the organic salt and the salt matrix in order to
account for the dramatic observations.   The effects are seen both
for mixtures deposited on surfaces from solution and those
burnished on as solids.   Further evidence of mixing is provided by
another effect of the matrix, the suppression of intermolecular
reactions.   Figure 3 shows SIMS spectra of carnitine hydrochloride
as the neat material and when diluted 200 times with NH_4Cl.   The
neat compound undergoes intermolecular methylation to form the
quaternary methyl ester, as evidenced by the presence of m/z 176
fourteen mass units above the intact cation at m/z 162.   Addition
of NH_4Cl causes a dramatic decrease in the abundance of the
methylated cation, which indicates that neighboring group
interactions are minimized in the presence of the matrix.   The fact
that this effect can be seen even for a physical mixture indicates
that ion-beam-induced mixing must occur, either due to the agency
of prior impact events or in conjunction with the desorbing impact
itself.   Impact of a single primary particle could cause mixing

over a much larger surface area than would be sampled in the concurrent sputtering event, so allowing mixing to occur under static SIMS conditions. Such mixing is consistent with observations by Michl (23) of cluster emission in SIMS of small molecule solids at cryogenic temperatures, and with the exchange of chlorine for hydrogen in physical mixtures of quaternary ammonium and metal chloride salts (24). Michl has advanced a cluster emission model which accounts for mixing in the hot ejected clusters and for subsequent cooling by ejection of small molecules. This is an attractive explanation of the effects we observe for room temperature salt matrices.

The mechanistic implications of another matrix effect currently under study (25) are not yet clear. This effect concerns the degree of fragmentation of anions as influenced by the nature of cations present in the matrix. Figure 4 illustrates negative ion SIMS spectra of potassium and ammonium perrhenate. A striking difference in the two spectra is the much greater abundance of ReO_3^- in the spectrum of NH_4ReO_4 vs. that of $KReO_4$. If a physical mixture of NH_4ReO_4 and KCl is studied, the SIMS spectrum appears identical to that of $KReO_4$. The suppression of fragmentation in NH_4ReO_4 is also effected by RbCl and CsCl, but not by NaCl or LiCl. This counterion effect may be associated with electron transfer phenomena in the selvedge.

While questions remain as to the origin of some matrix effects, manipulation of the matrix is being employed increasingly as a means for optimizing analyses. Chemistry plays a major role in processes occurring during sputtering. For example, Brønsted acid-base concepts have been used effectively to increase the abundances of certain ions in SIMS. Benninghoven (26) found that acidification of a solution from which glycine was deposited on a surface resulted in an increased abundance of [M+H]$^+$ ions, while addition of sodium hydroxide to the solution gave a maximum yield of [M-H]$^-$ ions. Addition of p-toluenesulfonic acid to sample matrices has been demonstrated to give enhancements in [M+H]$^+$ ion yields for a number of biological compounds in both SIMS and laser desorption (27). Recently this matrix enhancement strategy has

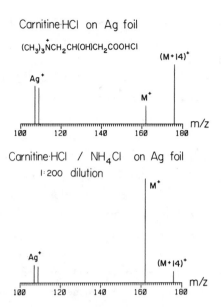

Figure 3. Decrease in abundance of the product of an intermolecular reaction (methylation to give 176[+]) upon dilution in a matrix.

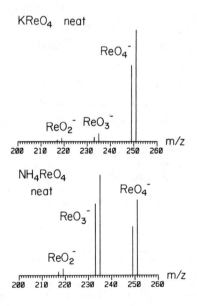

Figure 4. Effect of counterion upon fragmentation observed in SIMS.

been extended to the use of Lewis acids and bases. Todd, et
al. (28) have achieved enhancements in $M^{+\cdot}$ ion yields in SIMS
of pyrene dissolved in molten $SbCl_3$. The key here is the
electron-accepting ability of $SbCl_3$, possibly resulting in
formation of a charge transfer complex from which the pyrene
molecular ion is easily desorbed.

Other experiments. While matrix effects can provide insight
into the nature of the desorption process, there exist other more
direct methods for obtaining information on the SIMS mechanism.
Monitoring of secondary ion intensity vs. secondary ion or
primary ion kinetic energy can sometimes provide mechanistic
information (see, for example, Rabalais, et al. (29)).
Primary ion currents can be varied, and beam damage at high fluxes
may be indicative of particular processes (see below). Also of use
is variation of the nature of the primary ion itself. For example,
sputtering of metals or semiconductors with O_2^+ produces
enhancement in ion yields in comparison to inert gases such as
Ar^+; this can be explained as the result of oxidation of the
material, forming a band gap which limits neutralization of
sputtered ions on emission (30). (Oxidation also allows direct
desorption of precharged ions.) Variation of the azimuthal angle
of incidence of primary ions and measurement of the angular
distributions of secondary ions have been used by Winograd (31),
often in conjunction with classical dynamics calculations (32),
to study cluster formation and surface configurations of adsorbed
species.

The time dependence of desorption remains a little-explored
but potentially useful approach for mechanistic studies. Cotter
(33) has monitored secondary ion kinetic energies in a laser
desorption (LD) time-of-flight instrument. Laser pulses 40 ns wide
were used to desorb K^+ ions from solid KCl, and the ions were
sampled at variable times after the laser pulse. Emission persists
for several microseconds after excitation, and secondary ion
kinetic energies were found to decrease when examined at longer
times after excitation. This result supports a thermal model for

LD in which relaxation occurs after an initial high temperature spike. Time-of-flight instruments are well suited for such time-dependent studies, since the time variable is measured directly. In addition to its use in LD, TOF is used exclusively in plasma desorption (PD) (34) and is being used increasingly in SIMS (35,36). It is noteworthy that a time resolution of about 10^{-11} s in field ionization is the present limit attainable in any technique (37). If the desorption process itself occurs on the order of one vibrational period (10^{-13} s) (38), any information to be gained about the very early stages of desorption is at present not obtainable through measurement of time delays. Such small delays are of prime interest, however, since processes occurring through selvedge reactions could then be distinguished from direct emission processes, direct emission could in turn be distinguished from unimolecular fragmentation, and information could be gained on the nature of energy randomization at the surface.

There may also be mechanistic information available in TOF peak shapes. Figure 5 compares types of peaks encountered in three different analyzers if a repelling voltage V (39) is applied after full acceleration from the ion source, but prior to detection (TOF) or mass analysis (quadrupole and magnetic sector). For TOF and magnetic analyzers, broad metastable peaks are indicative of decomposition after full acceleration. Since the quadrupole analyzes mass directly, there is no metastable peak in the unmodified analyzer, regardless of the value of V. In the case of magnetic sector analysis, a metastable peak appears as long as V is held below a critical value V_c, where V_c equals the kinetic energy partitioned into the fragment ion. (This discussion neglects fragmentation in the region following application of V_c; this would give rise to a second peak.) When V surpasses V_c, the metastable peak is no longer observed, while the narrow peak due to ions with full accelerating voltage remains. If V is zero in the TOF case, the peak shows a pronounced broadening at the base. Application of a repelling voltage allows separation of this broad peak into two components: that due to stable and that due to

metastable ions. Increasing V to V_c and beyond results in
complete repulsion of the metastable component, with a slight
lengthening of the flight time for stable ions as a result of the
repelling voltage. Determination of the threshold energy for
metastable decomposition gives information on the internal energies
of secondary ions, which ultimately leads back to the desorption
process. Determination of internal energy distributions in
sputtered ions seems feasible using new methods (40), and this
should provide further mechanistic insights.

A time-of-flight instrument (41) has recently been
constructed for the purpose of observing metastable decompositions
of ions formed in plasma desorption. The instrument uses an
electrostatic mirror to reflect the ion beam while permitting
neutrals to pass through unhindered to a detector. One operating
mode involves acquisition of a spectrum with the electrostatic
mirror disabled, allowing both ions and neutrals to strike the
detector behind the mirror. A second measurement with the mirror
operating will give a spectrum of neutrals only. These neutrals
must have been formed by in-flight decomposition of ions, since
neutrals could not be accelerated out of the instrument source.
The neutrals, having the same velocities as the parent ions from
which they arise, appear at flight times identical to those of the
unfragmented parent ions. The integrated peak ratio
(neutrals)/(neutrals + ions) gives a direct measure of the
contribution of metastables to a TOF peak. In some cases, 80% to
90% of the peak area comes from metastable decompositions (42).
A second operating mode of this instrument is a coincidence
experiment in which detection of both neutrals and reflected ions
is accomplished simultaneously. It is possible in this mode to
identify fragment ions formed by dissociation of a particular
parent ion—a capability of great importance for ion structural and
mechanistic studies.

There are a number of recent studies which, though not aimed
specifically at determining the mechanism of desorption, point the
way to future experiments in this area. FAB of quinone antibiotics
(43) appears to be accompanied by reduction. Quinones with low

reduction potentials give large ($[M+2H]^{+\cdot}$ + $[M+3H]^{+}$)/$[M+H]^{+}$ and ($[M+H]^{-}$ + $M^{-\cdot}$)/$[M-H]^{-}$ ratios. The ratios follow polarographically-measured potentials and are dependent on solvent, analysis time, and quinone structure. A FAB study by Bursey, et al. (44) has quantitated the competition for alkali metal cations between solvent glycerol and solute phthalic acid. A fuller understanding of this competition and of the processes involved in the reductions could aid greatly in the explanation of matrix effects. The information now available points to significant contributions of both selvedge ion/molecule reactions and subsequent unimolecular dissociations. Using tandem mass spectrometry (see below), Keough (45) has found strong evidence for desolvation processes in FAB spectra of sucrose and glucose in NaCl-doped glycerol. Solvated clusters generated in FAB were collisionally dissociated and found to yield fragments of the same composition as other ions in the FAB survey spectrum. It is reasonable to suppose that many ions in FAB and SIMS result from similar unimolecular dissociations.

Control of target surface potential has been used to increase the specificity of the information available from SIMS. Detection of inorganic constituents in coal and polymer samples (46) is hampered by the strong molecular ion backgrounds often present. By charging a surface to a particular value, secondary organic ions with low kinetic energies can be suppressed, while the higher energy atomic ions are transmitted. This method is a variant on secondary ion kinetic energy analysis.

An experiment related to that above involves heating or cooling surfaces to observe the effects, if any, on sputtering. A SIMS study of ethylene adsorbed on ruthenium (47) suggests the type of information obtainable with surface temperature variation. In Figure 6, ions produced by sputtering were monitored during application of a temperature ramp to a Ru(001) surface. While all the catalytic implications of this data will not be discussed, the figure is informative in a mechanistic sense. For example, the fact that RuC^{+} and RuC_{2}^{+} increase as Ru^{+} increases indicates a local surface RuC_{2} unit or recombination of neutral C

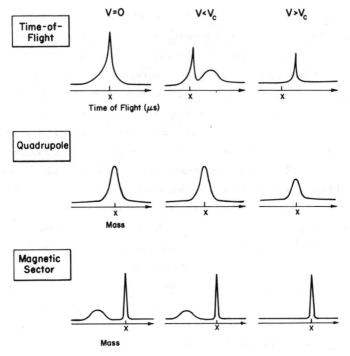

Figure 5. Effect of gas phase fragmentation upon peak shapes observed using different mass analyzers for SIMS.

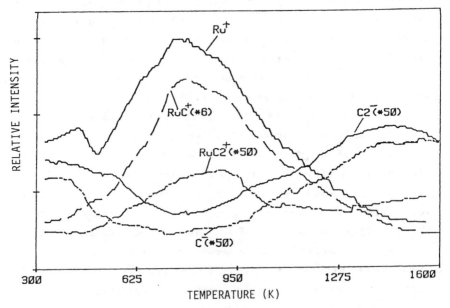

Figure 6. Surface temperature dependence of the SIMS spectrum of ethylene on Ru.

or C_2 species with Ru^+ in the selvedge. The fact that no positive C^+ or C_2^+ ions could be observed rules out participation by neutral Ru in the cationization process.

Alternative methods of energy deposition. The relationship between SIMS and other desorption ionization techniques can provide additional mechanistic insights. In laser desorption, as in SIMS, the processes of cationization/anionization, direct desorption of precharged materials, and electron ionization (Equations 1-3) are easily recognized. The similarities between the two methods can be illustrated by comparing studies of doubly-charged organic salts (such as diquaternary ammonium salts) (33,48). These compounds fail to show doubly-charged ions in LD, although abundant $[M-H]^+$ or $[M-CH_3]^+$ ions are seen along with singly-charged fragments. Dissociation of diquaternary ammonium salts into two singly-charged quaternary ions is sometimes observed, as is substitution of halogens for hydrogen. In rare instances, doubly-charged ions can be observed in SIMS (49). The similarities between SIMS and LD spectra, especially those obtained with pulsed beams, far outweigh the differences, and this is significant in view of the completely different means used for excitation.

The techniques of field desorption (FD) and electrohydrodynamic ionization (EHD) differ from SIMS, LD, and FAB in their physical basis and in features seen in the spectra. For example, the diquaternary ammonium salts discussed above yield intact doubly-charged ions, and fragmentation is less extensive. Nonetheless, the three classes of ions described for molecular SIMS are generally seen in FD (50). An EHD study of a series of diquaternary ammonium salts (51) led to the conclusion that the amount of internal energy deposited in EHD is less than in SIMS and FAB. The same study also indicated that FAB (liquid matrix) deposits less energy than SIMS (solid matrix), so in this case the order of energy deposition is SIMS > FAB > EHD.

It is generally recognized that FAB and SIMS really represent one and the same sputtering phenomenon. The charge borne by the primary particle is of little mechanistic consequence, and it is the liquid matrix which is resposible for the advantages and disadvantages of FAB. A direct comparison between spectra obtained with Cs^+ and Xe^o primary beams has been reported ($\underline{52}$). Peptides, lipids, nucleotides, and steroids desorbed from glycerol are found to exhibit virtually identical spectra with both Cs^+ and Xe^o primary particle bombardment. The only significant difference is an increase in secondary ion intensity with Cs^+, probably due to the ability to better focus the charged beam.

In contrast to SIMS and FAB, plasma desorption employs heavy, very high energy (MeV) fission fragments as primary particles. The flux of these particles is much lower than fluxes used in SIMS and FAB. The primary excitation process in PD is electronic, while that in SIMS and FAB is generally considered to arise from nuclear stopping ($\underline{53}$). Due to this electronic excitation there is a dependence on the charge state of the primary particle, unlike the situation in SIMS and FAB. Despite differences in the early stages of ionization, however, the same types of ions seen in SIMS, FAB, LD, and FD are also seen in PD. This suggests conversion of energy to a common form after the initial energization process.

Summary of mechanism. Taking into account the data in hand (and that which follows), it is possible to classify the many processes which occur in molecular SIMS in a qualitative manner. Table II, while not complete, illustrates many of the processes which contribute to molecular SIMS, FAB, and LD spectra. The following synopsis of the mechanism complements the material in Table II:

(1) Upon particle impact, energy is deposited into a surface and distributed through momentum transfer and vibrational and rotational excitation. This leads to heating, electron and photon emission, neutral particle emission, and ion emission.

(2) There are, initially, at least three types of molecular ions from which other ions may be derived. These are formed by electron ionization, direct desorption, and ion/molecule processes, especially cationization/anionization. The means by which these

Table II. Ion formation processes in molecular SIMS

Reaction	Probable region of occurrence	Comments
M^0 (s) $\longrightarrow$ M^0 (g) $\xrightarrow{+\ e^-}$ $M^{+\cdot}$ (g)	selvedge	electron ionization
C^+ (s) $\longrightarrow$ C^+ (g)	surface	direct desorption
M^0 (s) $\longrightarrow$ M^0 (g) $\xrightarrow{+\ C^+}$ $(M + C)^+$ (g)	selvedge	cationization
M^0 (s) $\longrightarrow$ M^0 (g) $\xrightarrow{+\ A^-}$ $(M + A)^-$ (g)	selvedge	anionization
M^0 (s) $\longrightarrow$ M^0 (g) $\xrightarrow{+\ H^+}$ $(M + H)^+$ (g)	selvedge	cationization by protonation
$(C_m A_n)^{\pm}$ (s) $\longrightarrow$ $(C_m A_n)^{\pm}$ (g)	surface	direct cluster desorption
C^+ (s) $+\ nS^0$ (l) $\longrightarrow$ CS_n^+ (g)	surface (liquid)	solvent attachment
C^+ (g) $+\ nS^0$ (g) $\longrightarrow$ CS_n^+ (g)	selvedge	cationization of solvent
CS_n^+ (g) $\longrightarrow$ C^+ (g) $+\ nS^0$ (g)	free vacuum	desolvation
M_1^0 (s) $+\ M_2^0$ (s,l) $\xrightarrow{\pm\ e^-}$ $M_1^{\pm}$ (g) $+\ M_2^{\mp}$ (s,l)	selvedge	electron transfer
M^0 (s) $\longrightarrow$ F (g) $\xrightarrow{+\ C_n^{\pm}\ (g)}$ $(C_n + F)^{\pm}$ (g)	surface $\longrightarrow$ selvedge	"beam damage"
$[C^0$ (s) $+ Ad$ (g)$]$ $\longrightarrow$ $(C + Ad)^{\pm}$ (g)	surface	intact emission
Ad (g) $\xrightarrow{+\ C^+}$ $(C + Ad)^+$ (g)	selvedge	cationization
nC (s) $\longrightarrow$ $C_n^{\pm}$ (g)	surface	intact cluster emission
M_1^0 (s) $+\ M_1^0$ (s) $\xrightarrow{+\ H^+}$ $(2M_1 + H)^+$ (g) $\longrightarrow$ M_2^+ (g)	surface	surface reaction
$\curvearrowleft$ F	free vacuum	unimolecular fragmentation

M = molecule, C^+ = cation, A^- = anion, C = atom or molecule, S = solvent, Ad = adsorbed species, F = fragment, m,n = positive integers ions arise are to some degree obscure. For example, electron ionization may involve direct selvedge ionization of neutrals, but it could also involve charge exchange or electron loss concurrently with the actual desorption of a molecule.

(3) Ions formed in the selvedge can react with other ions or neutrals to form clusters and other secondary products. Ions sputtered from liquid matrices can cationize solvent molecules.

(4) Some clusters may be desorbed as intact units. Likewise, solvated adducts may be desorbed from liquid matrices directly.

(5) Under normal experimental conditions, unimolecular fragmentation and desolvation occur predominantly in free vacuum above the selvedge.

(6) Especially under high flux conditions, some molecules can be fragmented directly at the surface, and these fragments can form adducts in the selvedge.

Although they specifically describe the behavior of organic molecules, it is interesting to examine the extent to which the preceding generalizations apply to simple metal salts (54). Figure 7 is a portion of the FAB spectrum of $SbCl_3$. A distinguishing feature of this spectrum is the presence of successive losses of neutral HCl from ions solvated by glycerol. The negative FAB spectrum of $SnCl_4$ contains similar HCl losses and a series of $SnCl_x^-$ peaks arising from addition of Cl^- to $SnCl_4$ (presumably in the selvedge) and apparently from successive losses of Cl_2 (free gas phase). In these cases, heterolysis of the substrate (e.g., to give $SnCl_5^-$ and $SnCl_3^+$) followed by solvation in the selvedge and unimolecular fragmentation via elimination of stable neutral molecules accounts for much of the behavior observed in FAB.

Tandem Mass Spectrometry (MS/MS)

An additional dimension of information can be added to desorption techniques if a second stage of mass analysis is used. Typically an ion is mass-selected with a quadrupole or magnetic sector and allowed to pass into a collision cell. A target gas in the

collision cell causes collisional activation of the ion, which then
fragments. Fragment ions are mass-analyzed to obtain a daughter
spectrum. An alternative scan mode--the parent scan--involves
setting the second analyzer to select ions of a constant mass while
the first analyzer is scanned; this produces a spectrum of ions
which fragment to give a particular daughter ion. MS/MS is very
useful in ion structure determinations, especially in cases where
an abundant molecular ion and an absence of fragment ions are
encountered. It also has proven valuable in providing evidence
that gas phase dissociations are responsible for the fragment ions
observed in SIMS and FAB spectra of (i) thiamine, (ii) NH_4Cl, and
(iii) silver propionate, to give a few particular examples which
will be considered in turn.

MS/MS daughter spectra of a number of ions generated by
sputtering of organic materials in liquid and solid matrices have
been reported (55). In one experiment the intact cation (m/z
265) of thiamine HCl was selected and found to fragment by cleavage
into two main fragments (m/z 122 and m/z 144). This reaction is
identical to that seen in the SIMS survey mass spectrum of thiamine
HCl itself. An ion at m/z 357 was observed in the SIMS spectrum of
thiamine HCl from glycerol; this is the adduct $[M+G]^+$, where M is
the intact cation and G is glycerol. The daughter spectrum of this
ion was expected to show simple loss of glycerol accompanied by
thiamine fragments. This indeed occurs, but the base peak in the
daughter spectrum is m/z 238, which formally corresponds to loss of
glycerol and an additional 27 mass units (probably HCN).
Apparently the glycerol solvent reacts with the thiamine cation to
form new covalent bonds. This is an example of an intimate
solvent-analyte association (as contrasted with solid matrix data
in which very loose associations are normally indicated).

In the same study, an unexpected result was observed in the
daughter spectrum of the intact cation (m/z 162) from carnitine
(see Matrix effects, above). The base peak in SIMS MS/MS is
m/z 103, corresponding to loss of trimethylamine by rearrangement,
while the same ion in the SIMS survey spectrum is nearly absent.
This is a striking case of different behavior between SIMS and gas

phase fragmentations. Its origins are not entirely clear, although dissociation in SIMS is influenced by the nature of the matrix, pointing to the occurrence of fragmentation in the selvedge.

A study of silver carboxylates (56) employing evaporation followed by electron ionization (EI) in conjunction with MS/MS allowed structures to be formulated for certain ions observed in SIMS. The SIMS spectrum of silver propionate, for example, contains unusual ions indicative of high energy bond cleavages, in contrast to most other molecular SIMS spectra. These ions are $AgCO^+$, $AgCO_2^+$, Ag_2O^+, and especially $Ag_2C_2H_5^+$ and $Ag_2propCH_2^+$ (prop = $CO_2C_2H_5$). In addition, Ag^+, Ag_2^+, Ag_3^+, Ag_2H^+, $Agprop^+$, and Ag_2prop^+ were observed. Ideally the structures of some of these ions would be determined with SIMS MS/MS. When this direct method was not available, EI spectra from a vaporized sample were acquired on a triple quadrupole mass spectrometer. The ions observed in EI were the same as in SIMS, with the exception of Ag_3^+ (absent) and Ag_2CO^+ and $Ag_2CO_2^+$ (very weak). The similarity between the spectra may be indicative of both (i) thermal processes occurring in SIMS and (ii) unimolecular dissociations occurring in the free vacuum region. Parent and daughter MS/MS scans produced the data in Scheme 1, which shows the origins and fates of the various Ag_2-containing species. Ag_2prop^+ is seen to be a parent of Ag_2CO^+ and $Ag_2CO_2^+$, which arise by cleavage of C-C bonds in the metal-containing cluster ion. Formation of these ions in SIMS is therefore reasonably ascribed to gas phase dissociations. Particularly interesting is the Ag_2H^+ ion, which does not fragment to Ag_2^+, in contrast to ions like Ag_2CO^+ and Ag_2prop^+, which do. The conclusion is that Ag_2CO^+ contains a Ag-Ag bond, while Ag_2H^+ has a proton-bound Ag-H^+-Ag structure. In cases like this, MS/MS can offer important insights into molecular structure otherwise unobtainable with only one stage of mass analysis.

The observation of high mass clusters is familiar in SIMS, and so it is not surprising that NH_4Cl clusters are observed in FAB (21). Like the solid alkali halides (57), NH_4Cl produces a

series of clusters containing gaps at certain masses. Figure 8 is
a portion of the positive ion FAB spectrum of NH_4Cl. Clusters
are clearly evident by the distinctive chlorine isotope patterns
and are described by the formula $[(NH_4)_x Cl_{x-1}]^+$. Clusters
in SIMS, too, commonly have this general $[C_x A_{x-1}]^+$ formula
(30). In this particular case, clusters were observed from x = 1
to x = 38. At increasingly high masses, ion intensities decrease,
with exceptions at x = 38, 32, 23, and 14, where intensities are
notably greater than expected. Gaps occur in the cluster series at
x = 31, 24, 22, and 15. The gap for x = 15
($[(NH_4)_{15} Cl_{14}]^+$) is clearly evident in Figure 8. To study
the absence of an x = 15 ion further, daughter spectra of the x =
17 (m/z 872) and x = 16 (m/z 819) ions were recorded. The x = 17
ion was found to fragment to x = 16, 14, 13, . . . , with x = 15
being absent. The x = 16 ion was found to lose two NH_4Cl groups
to give the series x = 14, 13, 12, ... Thus, the x = 15
cluster was not observed in any of the experiments and is clearly
of low stability. The MS/MS data confirm that the key elements of
the desorption ionization spectra arise as a result of unimolecular
fragmentations, as shown earlier for cesium iodide by Standing and
coworkers (58).

Applications

Quantitative and trace analysis. Quantitative analysis by
SIMS must be performed with an awareness that different materials
have different sputtering yields, and thus a large peak for one
material may or may not indicate that it is present in greater
abundance than a material which gives a smaller peak. If
appropriate standards are used, however, it is possible to
accurately quantitate materials in SIMS. As one example, full
spectra of good quality and signal persistence of at least thirty
minutes under static SIMS conditions have been reported for
therapeutic quaternary pyridine aldoximes at the 50 ng level
(59). Monitoring of only one ion and/or signal averaging could
be used to lower the detection limit substantially.

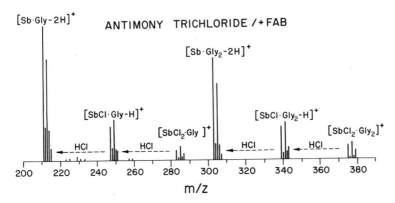

Figure 7. FAB spectrum of SbCl$_3$ in glycerol showing fragmentations leading to lower mass ions.

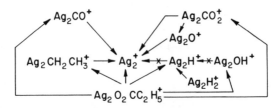

Scheme 1. Ion relationships in silver propionate determined by Ms/Ms. Reproduced with permission from Ref. 56. Copyright 1984, American Chemical Society.

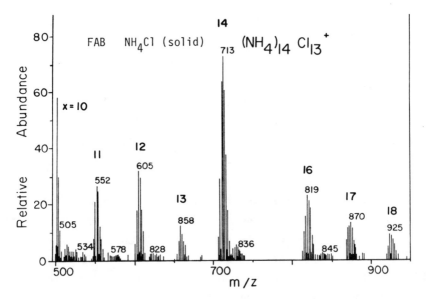

Figure 8. FAB spectrum of solid NH$_4$Cl showing gap in cluster ions at (NH$_4$)$_{15}$Cl$_{14}^+$.

As in many other quantitative mass spectrometric studies, the best quantitative results are often obtained in molecular SIMS through use of isotopically-labelled internal standards. This procedure has been successfully used in quantitating the extent of methylation of nucleosides produced by in vitro methylation of DNA (60). Figure 9A is the SIMS spectrum of 3-methylcytidine (m^3C), characterized by the 3-methylcytosine ion (arising from cleavage of the ribose sugar) at m/z 126. Methylated polycytidylic acid, enzymatically degraded in the presence of deuterated 3-methylcytidine ($[methyl-^2H_3]m^3C$), gives the SIMS spectrum in Figure 9B. The ion at m/z 112 is protonated cytosine. The deuterated standard shows a peak at m/z 129 which can be compared to the m/z 126 peak arising from m^3C produced by degradation. In this case, 22% of m^3C was found in the methylated polycytidylic acid, a figure which agrees with that obtained in studies by high performance liquid chromatography and nuclear magnetic resonance.

An important tool for quantitative as well as qualitative analysis by molecular SIMS is reverse derivatization (27). In contrast to derivatization normally encountered in mass spectrometry, in which an involatile material is made more volatile, derivatization reactions appropriate to DI yield a salt which can be efficiently ionized. Figure 10 demonstrates one such derivatization reaction (61) in which corticosterone reacts with Girard's reagent P to form a pyridinium salt. Positive ion FAB analysis of the derivatized material produces an abundant intact cation for as little as 1 μg of sample ketosteroid.

Trace analysis by SIMS can achieve strikingly low detection limits, even for impure samples and mixtures, when care is taken to optimize each aspect of the experiment. Benninghoven's work on biomolecules in the 1000-2000 molecular weight range (62) demonstrates this point, with 10^{-15} mol detection limits being available in a TOF analyzer. The important factor in this work is the use of a sputter-cleaned noble metal substrate, which allows deposition from solution of monolayers or submonolayers of sample. The use of chemical ionization with MS/MS has enabled detection of

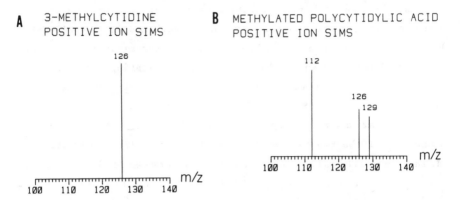

Figure 9. (A) Partial SIMS spectrum of 3-methylcytidine, with the 3-methylcytosine ion at m/z 126. (B) Quantitation of 3-methylcytosine (m/z 126) by SIMS using a d₃-labeled internal standard (m/z 129).

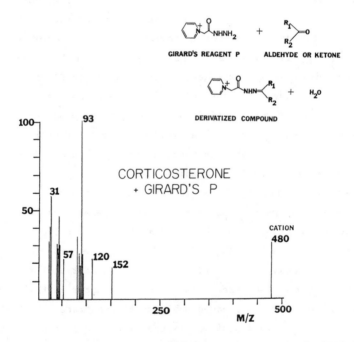

Figure 10. Derivatization in SIMS used to obtain efficient ionization of nonpolar compounds.

picogram amounts of components in biological mixtures (63), and SIMS may provide even lower detection limits for mixture analysis.

<u>Chromatography</u>. SIMS is particularly well-suited to the analysis of materials separated by paper chromatography, thin layer chromatography (TLC), and electrophoresis. Since sample is already present on a solid support in these methods, direct analysis of separated components is easily performed. Detection of the quaternary ammonium salt muscarine in a mushroom extract (64) exemplifies the technique. Detection limits are about 10 µg for <u>in situ</u> analysis of TLC spots, but by scraping off spots and redepositing them on metal supports, limits two to three orders of magnitude better are obtained. This is due to removal of much of the background signal which arises from the plate and to sampling of more material (SIMS samples only surface material, so material absorbed into the silica gel or cellulose layers will not be sampled). SIMS has also been used as a detector for liquid chromatography (65), but this is experimentally more difficult than detection of TLC spots.

<u>Ion Chemistry</u>. Quite often in molecular SIMS, fragmentation is observed which can be explained through simple schemes of the sort commonly encountered in electron impact and chemical ionization mass spectrometry. Consider Figure 11, the SIMS spectrum of nicotinamide (18). A number of peaks are apparent, and Scheme 2 serves to explain many of them. After argentation at the ring nitrogen atom, an additional silver atom may replace a hydrogen to give the $[2Ag+M-H]^+$ ion, followed by loss of HNCO to give $[2Ag+M-H-HNCO]^+$. (The attachment of Ag_2 to $[M-H]^+$ is apparently precluded here, since no $[M-H]^+$ fragment ion is seen.) Attachment of two silver atoms is probably a selvedge process, while loss of HNCO is a free gas phase fragmentation, as indicated by direct experiments with the corresponding protonated compound. Other peaks in the spectrum are easily recognizable: the radical nicotinamide cation $M^{+\cdot}$, the protonated nicotinamide cation $[M+H]^+$, the silver cluster Ag_3^+, and a Ag_2Cl^+ impurity cluster with the familiar $[C_xA_{x-1}]^+$ composition.

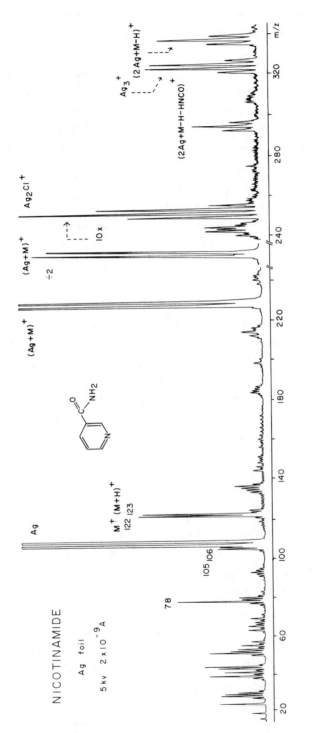

Figure 11. SIMS spectrum of nicotinamide. Reproduced with permission from Ref. 18. Copyright 1981, Elsevier Science Publishers B.V.

Interesting behavior has been observed in SIMS of some mixtures of tetrahedral inorganic complexes (66). Table III presents ions observed in the SIMS spectrum of a 1:1 mixture of copper methyl and silver cyclohexyl isocyanides. A graphite support and granulated graphite matrix were found to give the best signal/noise ratios. Considerable mixing in the selvedge is suggested by ligand exchange (e.g., $[Cu(CNCy)_2]^+$ and $[Ag(CNCH_3)_2]^+$) and by the presence of mixed ligand species ($[Cu(CNCH_3)(CNCy)]^+$). On the basis of a growing number of studies of inorganic compounds and coordination complexes (67-69) it is apparent that intermolecular reactions are far more likely for these compounds than for organic compounds. One surmises that the stronger intermolecular and weaker intramolecular bonds in inorganic compounds cause this difference in behavior from typical organic samples.

Catalysis. SIMS has been used to study systems of catalytic interest, including nickel complexes supported on insoluble polymeric and oxidic materials (70). Figure 12A is the SIMS spectrum of a nickel cyclopentadiene triphenylphosphine complex on silver. The nickel peaks are characteristic of the complex, while the $Ag(PPh_3)^+$ peaks indicate selvedge cationization of triphenylphosphine by sputtered Ag^+. If alumina is added to the sample matrix, the resulting SIMS spectrum (Figure 12B) is, surprisingly, devoid of Ni-containing ions. This provides evidence of a strong Ni-alumina interaction. The ion $Ag(PPh_3)^+$ remains abundant, suggesting that Ni-phosphine bonds are cleaved preferentially to Ni-support bonds.

Studies of adsorbate interactions on catalytically reactive surfaces can also be carried out with SIMS. In one such study (71, 72), isotopic mixing was used to determine that carbon and oxygen atoms in $RuCO^+$ sputtered from a CO-saturated Ru(001) surface do not undergo exchange in the SIMS process. The SIMS spectrum of a Ru(001) surface after exposure to $C^{18}O$ contains $RuC^{18}O^+$ and $Ru_2C^{18}O^+$ ions, and also RuC^+, $Ru^{18}O^+$, and Ru_2C^+ ions indicative of C-O cleavage. Figure 13 shows the SIMS spectrum of a mixture of ^{13}CO and $C^{18}O$ on Ru(001). In

Table III. Ions present in the SIMS spectrum of $[Cu(CNCH_3)_4]PF_6$ and $[Ag(CNCy)_4]ClO_4$ 1:1 with granulated graphite on graphite.   Cy = cyclohexyl

Ions seen in neat $[Cu(CNCH_3)_4]PF_6$	Ions seen only in the mixture	Ions seen in neat $[Ag(CNCy)_4]ClO_4$
$[Cu(CNCH_3)_4]^+$ (1)		
$[Cu(CNCH_3)_3]^+$ (1)		
$[Cu(CNCH_3)_2]^+$	$[Cu(CNCy)_2]^+$	$[Ag(CNCy)_2]^+$
	$[Ag(CNCH_3)_2]^+$	
	$[Cu(CNCH_3)(CNCy)]^+$	
$[Cu(CNCH_3)_2+HCN]^+$	$[Ag(CNCH_3)_2+HCN]^+$ (2)	
$[Cu(CNCH_3)+HCN]^+$	$[Cu(CNCy)+HCN]^+$	$[Ag(CNCy)+HCN]^+$
$[Cu(CNCH_3)]^+$	$[Cu(CNCy)]^+$ (3)	$[Ag(CNCy)]^+$
$[Cu+2HCN]^+$		$[Ag+2HCN]^+$
$[Cu+HCN]^+$		$[Ag+HCN]^+$
Cu^+		Ag^+
		$C_6H_{11}^+$ (Cy)

(1) Present in the neat compound, but not in the mixture

(2) Presence uncertain due to overlap with $[Ag(CNCy)]^+$

(3) Presence uncertain due to overlap with $[Cu(CNCH_3)_2+HCN]^+$

C=O
NH$_2$
N+
Ag
+Ag·
-H·
(Ag+M)$^+$
+Ag·
-H·
Ag
C=O
NH-Ag
N+
Ag
C=O
NH$_2$
N+
Ag
-HNCO
(2Ag+M-H)$^+$
-HNCO
Ag
Ag
N+
Ag
N+
Ag
(2Ag+M-H-HNCO)$^+$

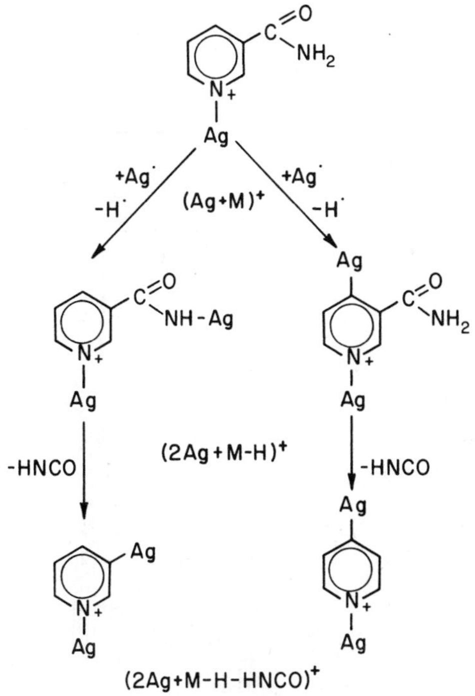

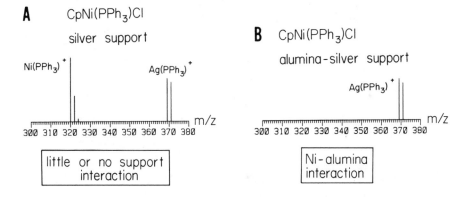

Scheme 2. SIMS fragmentation of nicotinamide on silver. Reproduced with permission from Ref. 18. Copyright 1981, Elsevier Science Publishers B.V.

A
CpNi(PPh$_3$)Cl
silver support
Ni(PPh$_3$)$^+$
Ag(PPh$_3$)$^+$
m/z
300 310 320 330 340 350 360 370 380
little or no support interaction
B
CpNi(PPh$_3$)Cl
alumina-silver support
Ag(PPh$_3$)$^+$
m/z
300 310 320 330 340 350 360 370 380
Ni-alumina interaction

Figure 12. (A) SIMS spectrum of a nickel complex showing characteristic ions (Cp = C_5H_5). (B) SIMS spectrum of the same complex in an alumina matrix.

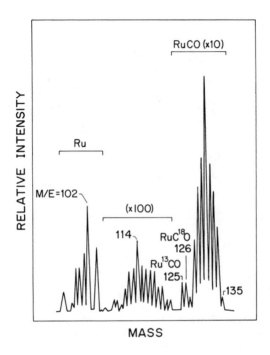

Figure 13. Isotopic labeling used to follow CO/Ru interactions
by SIMS. Reproduced with permission from Ref. 71. Copyright
1982, The American Physical Society.

spite of the multiple RuCO isotope peaks in the region of interest, the lack of a peak at m/z 124 (which corresponds to $^{96}Ru^{12}C^{16}O^+$) clearly indicates that atomic mixing does not occur in this case. The small peak at m/z 135 is $^{104}Ru^{13}C^{18}O^+$ formed from trace $^{13}C^{18}O$ impurities and not by mixing. Hence, although small amounts of C^+ and O^+ are formed during ion bombardment, RuCO$^+$ is composed of intact CO groups, a result which supports intact emission of the molecule.

Imaging. Although ion beam imaging of atomic species is a well-developed method in inorganic SIMS (73, 74), imaging of polyatomic secondary ions, discussed on a number of occasions (1,75), has only recently been demonstrated (76,77). The former study employed an Ar$^+$ ion gun with a 30μm spot size to image polymer films. In the latter experiment a vacuum pump fluid, the perfluoro polyether fomblin, was deposited on the wires of a 500 line/inch nickel mesh. The primary ion source was a liquid metal ion gun (78) capable of producing a spot size less than 0.3 μm in diameter. Mass analysis was performed with a high resolution double focusing mass spectrometer. Fomblin ions at m/z 131 and m/z 1281 were monitored to obtain two separate images, and in both images the supporting grid structure is clearly visible. Since signal at low mass was considerably stronger than that at high mass, the image quality for m/z 131 is much better than for m/z 1281. Nevertheless, high mass imaging was successful, and, as previously indicated (1, 75), one can envision future developments leading to in situ monitoring of biological materials.

Developing Areas

In addition to molecular imaging and such obvious advances as increased mass range and higher transmission analyzers, there are a number of new topics in molecular SIMS emerging as areas of interest. Not least of these is the study of chemical reactions at surfaces, which may be one source of the beam damage seen at high primary ion fluxes. A spectacular example of ion-beam-induced

reaction is seen in the thiophene-Ag system (79). If thiophene
vapor is introduced into the vacuum system of a SIMS instrument
concurrently with ion bombardment of a Ag surface, two unexpected
ion series are observed (in some experiments). One of these series
may be described by the general formula $[Ag+M+4H-n(14)]^+$, where M
is thiophene and n equals 0, 1, 2, or 3. The cationized reduction
product $[Ag+M+4H]^+$ is observed (n = 0); the expected $[Ag+M]^+$
adduct is not seen. Furthermore, for n = 1, 2, or 3, successive
losses of CH_2 from the hydrogenated adduct are indicated. There
are two processes occurring: complete hydrogenation (indicated by
the addition of four hydrogens to the silver-cationized molecule)
and hydrogenolysis (losses of CH_2). The presence of a large
Ag_2HS^+ ion is evidence that extensive decomposition of
thiophene is occurring and that the source of hydrogen is probably
thiophene itself. This behavior was observed in three cases out of
six, which implies a dependence on sample preparation. In the
unsuccessful cases, the simple $[Ag+M]^+$ adduct was seen, with no
contributions from hydrogenation or hydrogenolysis. Copper and
platinum were also used as substrates in these experiments, but
only $[Cu+M]^+$ and $[Pt+M]^+$ were observed.

Extensive C-C bond scission and rearrangements have been
observed in systems other than thiophene-Ag. Table IV presents a
number of such cases. The silver carboxylates (56) have already
been mentioned briefly. A preponderance of CO_2-containing adducts
suggests a gas phase analogue to the classical Hünsdiecker
reaction (Equation 4). The behavior observed with methyl acetate

$$RCO_2Ag + Br_2 \xrightarrow[HgO]{\Delta} RBr + CO_2 + AgBr \qquad (4)$$

vapor introduced over Cu foil may parallel the case of
thiophene-Ag. In addition, dosing of metals with O_2 in a
pretreatment chamber prior to SIMS analysis (80) also results in
the formation of unexpected ions during sputtering. It should be
emphasized that none of these data were obtained under
extraordinarily harsh conditions; maximum primary ion current
densities were on the order of 1×10^{-8} A cm^{-2}. The extensive
C-C cleavage, arising from implicitly high energy processes,
differs from earlier molecular SIMS data (1) in which simple

Table IV. Summary of C-C scission in molecular SIMS

Compound and Conditions	Important Ions
thiophene vapor/Ag foil	$[Ag + M + 4H - n(14)]^+$
(M = thiophene)	$[2Ag + M + 5H - n(14)]^+$
	$n = 0,1,2,3$
silver acetate on graphite	$AgCO^+$ or $AgC_2H_4^+$,
(OAc = acetate)	$AgCO_2^+$, $AgOAc^+$, Ag_2OAc^+
	$Ag_2O_2COAc^+$, $Ag_2O_2COAcCH_2^+$
silver propionate on graphite	$AgCO^+$ or $AgC_2H_4^+$,
(prop = propionate)	$AgCO_2^+$, $Agprop^+$, Ag_2H^+,
	Ag_2prop^+, $Ag_2O_2Cprop^+$,
	$Ag_2O_2CpropCH_2^+$
silver benzoate on graphite	$benzCO^+$ or $benzC_2H_4^+$,
(benz = benzoate)	$Agbenz^+$, $Agbenz_2^+$, Ag_2H^+
	Ag_2benz^+, $Ag_2O_2Cbenz^+$
methyl acetate vapor/Cu foil	$CuOH^+$, $CuCO^+$ or $CuC_2H_4^+$,
	$CuCO_2^+$, $CuC_2H_4CO^+$, Cu_2H^+,
	$CuMeOAc^+$, Cu_2O^+, Cu_2OH^+,
	$Cu_2C_2H_2^+$
methyl acetate vapor/Ag foil	$AgCO^+$ or $AgC_2H_4^+$, Ag_2H^+
copper acetate on graphite	$CuCO^+$ or $CuC_2H_4^+$, $CuCO_2^+$,
	Cu_2^+, Cu_2H^+, $Cu_2CH_3^+$,
	Cu_2OAc^+, Cu_3^+
O_2 pretreatment; CH_3CHO on Ag foil at $-115^{\circ}C$	$AgCO_2^+$, $Ag_2CH_3^+$
O_2 pretreatment; CH_3CHO on Ag powder at $25^{\circ}C$	$AgCH_3^+$, AgO^+, $AgCO^+$ or $AgC_2H_4^+$, $AgCO_2^+$, $Ag_2CH_3^+$, Ag_2O^+
O_2 pretreatment; C_2H_2 on Ag foil at $-115^{\circ}C$	$Ag_2C_2H^+$
O_2 pretreatment; C_2H_2 on Cu foil at $25^{\circ}C$	$Cu_3C_2^+$, $Cu_2C_2^+$, $Cu_2C_2H_2^+$

molecular species and relatively low energy fragmentations were apparent. Analytically, C-C scission is something to be avoided, but from a mechanistic viewpoint it may offer much insight. It is likely that at least some of these scissions occur at the surface, followed by cationization and combination in the selvedge. This is in contrast to the more common low energy unimolecular fragmentation which occurs in the free gas phase (vacuum). Strong chemisorption is one factor likely to favor surface bond scission.

Another area in which activity is expected to increase in the future involves the use of ion beams for modifying and/or synthesizing new materials. Ion implantation (81) has been used for some time in the semiconductor field, and sputtering of atomic materials (82) has a long history as a means for depositing uniform layers on surfaces. There have been no reports, however, of the sputter deposition of molecular species. A device has been constructed (83) for the purpose of sputtering organic materials, collecting them, and analyzing them by SIMS. Figure 14 illustrates the apparatus, which has been added to the sample preparation chamber of a commercial SIMS instrument. An ion gun is used to bombard a surface with a μA beam of 3keV Ar$^+$. This surface--the primary surface--may be a metal or graphite foil and may have material deposited on it. Secondary ions sputtered from the primary surface are collected at the secondary surface, which is mounted at the end of a magnetic transfer rod for in vacuo transfer into the SIMS main chamber after sputtering. The secondary surface, like the primary surface, can be any vacuum-compatible material. A potential may be applied between the two surfaces to control the energy of the secondary ions.

In an early version of this apparatus, the secondary surface is 2 cm away from the primary surface; hence, sputtered neutrals, which may constitute the majority of the transferred particles, present a problem during materials transfer. Despite the pitfalls of this simple design, transfers of material have been made which show a dependence on secondary ion energy. For Ag sputtered on phenanthroline-Pt, a large [Ag+M]$^+$ peak was seen for 300 eV Ag ions, while for 10 eV Ag ions the [Ag+M]$^+$ peak was much smaller.

The next step in the experiment will be to incorporate mass analysis of material sputtered from the primary surface in order to reject neutrals and to be more selective in what is deposited on the secondary surface. It is hoped that catalytically useful materials, such as mass-selected small metal clusters, may eventually be deposited on surfaces. Furthermore, it may be possible to transfer reactive organic species (such as those in Table IV) to create new materials through control of the potential between the two surfaces.

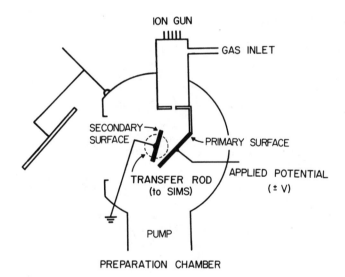

Figure 14. Apparatus to allow ion beams to be used in surface preparation. Reproduced with permission from Ref. 83. Copyright 1983, Elsevier Science Publishers B.V.

Acknowledgments

We thank Drs. Ken Busch, Steve Unger, and Tom Keough for
unpublished material. SJP thanks the Dow Chemical Corporation for
a graduate fellowship through the Purdue Industrial Associates
Program. The support of the National Science Foundation (CHE
8114410) is gratefully acknowledged.

Literature Cited

1. Day, R.J.; Unger, S.E.; Cooks, R.G. Anal. Chem. 1980, 52,
    557A.
2. Honda, F.; Lancaster, G.M.; Fukuda, Y.; Rabalais, J.W. J.
    Chem. Phys. 1978, 69, 4931.
3. Cooks, R.G.; Busch, K.L. Int. J. Mass Spectrom. Ion.
    Phys. 1983, 53, 111.
4. Day, R.J.; Unger, S.E.; Cooks, R.G. J. Am. Chem. Soc.
    1979, 101, 501.
5. Grade, H.; Winograd, N.; Cooks, R.G. J. Am. Chem. Soc.
    1977, 99, 7725.
6. Sigmund, P. Phys. Rev. 1969, 184, 383.
7. McCracken, G.M. Rep. Prog. Phys. 1975, 38, 241.
8. Benninghoven, A. In "Ion Formation from Organic Solids";
    Benninghoven, A., Ed.; SPRINGER SERIES IN CHEMICAL PHYSICS No.
    25, Springer-Verlag: Berlin, 1983; p. 64.
9. Magee, C.W. Int. J. Mass Spectrom. Ion Phys. 1983,
    49, 211.
10. Garrison, B.J. Ch.2 in this volume.
11. Winograd, N. Ch.5 in this volume.
12. Macfarlane, R.D. Ch.3 in this volume.
13. Hillenkamp, F. Ch.4 in this volume.
14. Garrison, B.J. J. Am. Chem. Soc. 1982, 104, 6211.
15. Moon, D.W.; Winograd, N. Int. J. Mass Spectrom. Ion
    Phys. 1983, 52, 217.
16. O'Hanlon, J.F. "A User's Guide to Vacuum Technology"; John
    Wiley & Sons: New York, 1980; p. 11.
17. Wong, S.S.; Röllgen, F.W.; Manz, I.; Przybylski, M.
    Biomed. Mass Spectrom. 1985, 12, 43.
18. Unger, S.E.; Day, R.J.; Cooks, R.G. Int. J. Mass
    Spectrom. Ion Phys. 1981, 39, 231.
19. Busch, K.L.; Hsu, B.H.; Xie, Y.-X.; Cooks, R.G. Anal.
    Chem. 1983, 55, 1157.
20. Barber, M.; Bordoli, R.S.; Elliot, G.J.; Sedgwick, R.D.;
    Tyler, A.N. Anal. Chem. 1982, 54, 645A.
21. Cooks, R.G.; Unger, S.E., unpublished data.
22. Hsu, B.H.; Xie, Y.-X.; Busch, K.L.; Cooks, R.G. Int. J.
    Mass Spectrom. Ion Phys. 1983, 51, 225.
23. Michl, J. Int. J. Mass Spectrom. Ion Phys. 1983,
    53, 255.

24. Day, R.J. Ph.D. Thesis, Purdue University, West Lafayette, Indiana, 1980.
25. Scheifers, S.M., unpublished data.
26. Benninghoven, A.; Sichtermann, W.K. Anal. Chem. 1978, 50, 1180.
27. Busch, K.L.; Unger, S.E.; Vincze, A.; Cooks, R.G.; Keough, T. J. Am. Chem. Soc. 1982, 104, 1507.
28. Groenewold, G.S.; Todd, P.J.; Buchanan, M.V. Anal. Chem. 1984, 56, 2251.
29. Lancaster, G.M.; Honda, F.; Fukuda, Y.; Rabalais, J.W. J. Am. Chem. Soc. 1979, 101, 1951.
30. Murray, P.T.; Rabalais, J.W. J. Am. Chem. Soc. 1981, 103, 1007.
31. Gibbs, R.A.; Winograd, N. Rev. Sci. Instrum. 1981, 52, 1148.
32. Foley, K.E.; Winograd, N.; Garrison, B.J.; Harrison, Jr., D.E. J. Chem. Phys. 1984, 80, 5254.
33. Cotter, R.J.; Tabet, J.-C. Int. J. Mass Spectrom. Ion Phys. 1983, 53, 151.
34. Macfarlane, R.D. Anal. Chem. 1983, 55, 1247A.
35. Chait, B.T.; Standing, K.G. Int. J. Mass Spectrom. Ion Phys. 1981, 40, 185.
36. Steffens, P.; Niehuis, E.; Friese, T.; Benninghoven, A. In "Ion Formation from Organic Solids"; Benninghoven, A., Ed.; SPRINGER SERIES IN CHEMICAL PHYSICS No. 25, Springer-Verlag: Berlin, 1983; p. 111.
37. Levsen, K. "Fundamental Aspects of Organic Mass Spectrometry"; Verlag Chemie: Weinheim, West Germany, 1978; p. 187.
38. Garrison, B.J. Int. J. Mass Spectrom. Ion Phys. 1983, 53, 243.
39. Chait, B.T.; Field, F.H. Int. J. Mass Spectrom. Ion Phys. 1981, 41, 17.
40. Kenttämaa, H.I.; Cooks, R.G. Int. J. Mass Spectrom. Ion Processes, in press.
41. Della Negra, S.; Le Beyec, Y. Int. J. Mass Spectrom. Ion Processes 1984, 61, 21.
42. Della Negra, S.; Le Beyec, Y. Institut de Physique Nucleaire, Université Paris-Sud, Report No. IPNO-DRE-85-01, 1985.
43. Cooper, R.; Unger, S. J. Antibiot. 1985, 38, 24.
44. Bursey, M.M.; Marbury, G.D.; Hass, J.R. Biomed. Mass Spectrom. 1984, 11, 522.
45. Keough, T., submitted for publication.
46. McIntyre, N.S.; Chauvin, W.J.; Martin, R.R. Anal. Chem. 1984, 56, 1519.
47. Lauderback, L.L. Ph.D. Thesis, Purdue University, West Lafayette, Indiana, 1982.
48. Hercules, D.M.; Day, R.J.; Balasanmugam, K.; Dang, T.A.; Li, C.P. Anal. Chem. 1982, 54, 280A.
49. Ryan, T.M.; Day, R.J.; Cooks, R.G. Anal. Chem. 1980, 52, 2054.
50. Röllgen, F.W. In "Ion Formation from Organic Solids"; Benninghoven, A., Ed.; SPRINGER SERIES IN CHEMICAL PHYSICS No. 25, Springer-Verlag: Berlin, 1983; p. 2.
51. Cook, K.D.; Chan, K.W.S. Int. J. Mass Spectrom. Ion Processes 1983, 54, 135.

52. Aberth, W.; Straub, K.M.; Burlingame, A.L. Anal. Chem. 1982, 54, 2029.
53. Macfarlane, R.D. Acc. Chem. Res. 1982, 15, 268.
54. Emary, W.B., unpublished data.
55. Glish, G.L.; Todd, P.J.; Busch, K.L.; Cooks, R.G. Int. J. Mass Spectrom. Ion Processes 1984, 56, 177.
56. Busch, K.L.; Cooks, R.G.; Walton, R.A.; Wood, K.V. Inorg. Chem. 1984, 23, 4093.
57. Barlak, T.M.; Wyatt, J.R.; Colton, R.J.; DeCorpo, J.J.; Campana, J.E. J. Am. Chem. Soc. 1982, 104, 1212.
58. Ens, W.; Beavis, R.; Standing, K.G. Phys. Rev. Lett. 1983, 50, 27.
59. Vincze, A.; Busch, K.L.; Cooks, R.G. Anal. Chim. Acta 1982, 136, 143.
60. Ashworth, D.J.; Chang, C.; Unger, S.E.; Cooks, R.G. J. Org. Chem. 1981, 46, 4770.
61. DiDonato, G.C.; Busch, K.L., submitted for publication.
62. Benninghoven, A.; Niehuis, E.; Friese, T.; Greifendorf, D.; Steffens, P. Org. Mass Spectrom. 1984, 19, 346.
63. Isern-Flecha, I., unpublished data.
64. Unger, S.E.; Vincze, A.; Cooks, R.G.; Chrisman, R.; Rothman, L.D. Anal. Chem. 1981, 53, 976.
65. Benninghoven, A.; Eicke, A.; Junack, M.; Sichtermann, W.; Krizek, J.; Peters, H. Org. Mass Spectrom. 1980, 15, 459.
66. Detter, L.D.; Walton, R.A.; Cooks, R.G., unpublished data.
67. Pierce, J.L.; Busch, K.L.; Walton, R.A.; Cooks, R.G. J. Am. Chem. Soc. 1981, 103, 2583.
68. Pierce, J.L.; Busch, K.L.; Cooks, R.G.; Walton, R.A. Inorg. Chem. 1982, 21, 2597.
69. Pierce, J.L.; Busch, K.L.; Cooks, R.G.; Walton, R.A. Inorg. Chem. 1983, 22, 2492.
70. Pierce, J.L.; Walton, R.A. J. Catal. 1983, 81, 375.
71. Lauderback, L.L.; Delgass, W.N. Phys. Rev. B 1982, 26, 5258.
72. Winograd, N.; Garrison, B.J.; Harrison, Jr., D.E. J. Chem. Phys. 1980, 73, 3473.
73. Brown, A.; Vickerman, J.C. Analyst 1984, 109, 851.
74. Brenna, J.T.; Morrison, G.H. Anal. Chem. 1984, 56, 2791.
75. Busch, K.L.; Cooks, R.G. Science 1982, 218, 247.
76. Briggs, D. Surf. Interface Anal. 1983, 5, 113.
77. Stoll, R.G.; Harvan, D.J.; Cole, R.B.; Hass, J.R. Proc. 32nd Annual Conference on Mass Spectrometry and Allied Topics, 1984, p. 838.
78. Barofsky, D. Ch.7 in this volume.
79. Unger, S.E.; Cooks, R.G.; Steinmetz, B.J.; Delgass, W.N. Surf. Sci. 1982, 116, L211.
80. Myers, T.E. Ph.D. Thesis, Purdue University, West Lafayette, Indiana, 1984.
81. Picraux, S.T.; Pope, L.E. Science 1984, 226, 615.
82. Greene, J.E. CRC Crit. Rev. Solid State Mater. Sci. 1983, 11, 47.
83. LaPack, M.A.; Pachuta, S.J.; Busch, K.L.; Cooks, R.G. Int. J. Mass Spectrom. Ion Phys. 1983, 53, 323.

RECEIVED April 16, 1985

Particle Bombardment as Viewed by Molecular Dynamics

Barbara J. Garrison

Department of Chemistry, The Pennsylvania State University, University Park, PA 16802

A classical dynamics model is used to investigate nuclear motion in solids due to bombardment by energetic atoms and ions. Of interest are the mechanisms of ejection and cluster formation both of elemental species such as Ni_n and Ar_n and molecular species where we have predicted intact ejection of benzene–C_6H_6, pyridine–C_5H_5N, napthalene–$C_{10}H_8$, biphenyl–$C_{12}H_{10}$ and coronene–$C_{24}H_{12}$. The results presented here show that the energy distributions of the parent molecular species, e.g. benzene, are narrower than those of atomic species, even though the ejection processes in both cases arise from energetic nuclear collisions. The bonding geometry also influences the ejection yield and angular distribution. The specific case of π-bonded and σ-bonded pyridine on a metal surface is discussed with comparisons between the calculated results and experimental data.

The bombardment of solids by energetic particle beams has attracted interest due to the ejection of large and novel species. These species can be molecules that are present in the original sample such as a dodecanucleotide (1) or clusters that are formed during the bombardment event, for example $[NO(N_2O_3)_3]^+$ ejected from solid nitrous oxide (2). Numerous other examples appear in these proceedings.

Our goal has been to understand the ejection mechanisms and the relationship of the clusters to the original configuration of atoms in the sample. Many mechanisms involving both the motion of the atomic nuclei and/or of electrons can be proposed to be responsible for ejecting the molecules. However, if a solid (or liquid) sample is bombarded by a heavy particle with energy in the 100–10000 eV range there must be energetic collisions between the atomic nuclei.

0097–6156/85/0291–0043$06.00/0

Thus as a starting point for understanding the bombardment process we have developed a classical dynamics procedure to model the motion of atomic nuclei. The predictions of the classical model for the observables can be compared to the data from sputtering, spectrometry (SIMS), fast atom bombardment mass spectrometry (FABMS), and plasma desorption mass spectrometry (PDMS) experiments. In the circumstances where there is favorable agreement between the results from the classical model and experimental data it can be concluded that collision cascades are important. The classical model then can be used to look at the microscopic processes which are not accessible from experiments in order to give us further insight into the ejection mechanisms.

Briefly, the theoretical model consists of approximating the solid and possible adsorbed molecules by a finite array of atoms (3-12). Assuming a pairwise interaction potential among all the atoms, Hamilton's equations of motion are integrated to yield the positions and momenta of all particles as a function of time during the collision cascade. The final positions and momenta can be used to determine the experimental observables such as total yield of ejected particles, energy distributions, angular distributions and possible cluster formation. One advantage of the classical procedure is that one can monitor the collision events and analyze microscopic mechanisms of various processes.

Mechanisms of Cluster Formation

From the classical dynamical treatment, it is possible to examine the cluster formation mechanism in detail and to provide semiquantitative information about cluster yields. In general, these calculations suggest that there are three basic mechanisms of cluster formation (12). First, for systems with atomic identity such as metals, or atomic adsorbates on a solid, the ejected atoms can interact with each other in the near-surface region above the crystal to form a cluster by a recombination type of process (3-5). This description would apply to clusters of the type M_nO_m observed in many types of SIMS experiments. In this case the atoms in the cluster do not need to arise from contiguous sites on the surface, although in the absence of long-range ionic forces the calculations indicate that most of them originate from a circular region of radius ~ 5 angstroms. This rearrangement, however, complicates any straightforward deduction of the surface structure from the composition of the observed clusters. We have observed an Ar_{25} cluster to eject from solid argon via this mechanism (13). We would also speculate that the alkali halide clusters $(CsI)_nCs^+$ with n as large as 70 (14) also form by this basic mechanism.

A second type of cluster emission involves molecular species which can be as simple as carbon monoxide or as complicated as the dodecanucleotide mentioned above. In the first case, the CO bond strength is ~ 11 eV, but the interaction with the surface is only about 1 eV. Calculations indicate that this energy difference is sufficient to allow ejection of CO molecules, although ~ 15 percent of them can be dissociated by the ion beam or by energetic metal atoms (6). For such molecular systems it is easy to infer the original atomic configurations of the molecule and to determine the surface chemical state. If CO were dissociated into oxygen and carbon atoms, for example, the calculations suggest that the amount of CO observed should drop dramatically.

Although the basic principles behind this intact ejection mechanism can be illustrated with carbon monoxide, the extrapolation to large bioorganic molecules is not necessarily obvious. Calculations have been performed for a series of organic molecules adsorbed on a Ni(001) surface to understand the mechanisms of molecular ejection (8-12). The first molecules which have more than just a few atoms examined are benzene which π-bonds on a metal surface and pyridine which can either π-bond or σ-bond on a metal surface. Larger structures, whose sizes approach the diameter of bioorganic molecules, are naphthalene, biphenyl and coronene whose adsorption structures are unknown. All the molecules except pyridine are assumed to π-bond on the surface.

In all cases we find that the molecular species may be ejected intact. From our theoretical calculations, three factors favor this process (8-9). First, a large molecule has many internal degrees of freedom and can absorb energy from an energetic collision without dissociating. Second, in the more massive framework of a large organic molecule, individual atoms will be small in size compared to a metal atom; thus, it is possible to strike several parts of the molecule in a concerted manner so that the entire molecule moves in one direction. Finally, by the time the organic molecule is struck, the energy of the primary particle has been dissipated so that the kinetic energies are tens of eVs rather than hundreds or thousands of eVs. These three factors are equally valid for the ejection of either carbon monoxide, benzene or coronene. However, in the cases of the larger molecules, we found that often 2-3 metal atoms would strike different parts of the molecule during the ejection process. The time for the molecules to eject after the primary particle has hit the sample is less than 200 femtoseconds (fs; 1fs=1x10^{-15}s). This intact ejection mechanism for molecules can be applied to molecular solids. We find for the bombardment of ice shows that the water molecules also eject intact (15).

It is difficult to make quantitative determinations of the fragment yields because the forces that govern all the rearrangement channels are not known. However, there is one interesting feature related to fragmentation that we have observed. Most of the fragments formed from direct collisions within ~ 0.2 ps are the parent molecule minus an H, CH, or C_2H_2. These arise from an energetic collision that rips off part of the molecule. In the case of biphenyl however, a severing between the two rings is observed to occur with some frequency. Thus the structure of the molecule influences the nature of the direct fragmentation process. These small CH type species will undoubtedly be formed during the ion bombardment process. To be detected, however, in a conventional SIMS or FABMS apparatus they must be formed as an ion. Within this classical model we are unable to predict the charge fraction.

The final mechanism for cluster ejection is essentially a hybrid mechanism involving both intact ejection and recombination. In the case of CO on Ni$_3$Fe, we find that the observed NiCO, Ni$_2$CO

and NiFeCO clusters form by a recombination of ejected Ni and Fe
atoms with ejected CO molecules. There is apparently no direct
relation between these moieties and linear and bridge-bond surface
states. In the case of cationized species such as $NiC_6H_6^+$ ions, we
propose a reaction of the type

$$Ni^+ + C_6H_6 \xrightarrow[surface]{} NiC_6H_6^+ \qquad\qquad (1)$$

The presumption that the Ni supplies the charge is based on the
fact that no $C_6H_6^+$ is observed (16) and that the ionization
potential of Ni is less than that of benzene.
     This final hybrid mechanism may be responsible for the
formation of the dimer ion of the dodecanucleotide (1) or of water
clusters (17). Each molecular unit ejects intact and then interacts
with other molecules in the near surface region to form the cluster
entities. In the case of $(H_2O)_2$ clusters our calculations indicate
that the two H_2O molecules originate from mostly adjacent sites on
the surface (15). This is a consequence of the relatively weak
H_2O-H_2O interaction. Ionic clusters such as $(H_2O)H^+$, however, can
form from an H_2O molecule and an H^+ ion that were further apart on
the surface.
     The fact that the composition of the ejected clusters may be
different from the original arrangement of surface atoms is
somewhat discouraging. As it turns out, however, there are
situations where the precise nature of the rearrangement can be
predicted theoretically. One example involves the measured
O_2^-/O^- ratio as a function of oxygen coverage on Ni(001). This
ratio is four times higher for 50 percent oxygen coverage
[c(2x2)coverage] than for 25 percent oxygen coverage
[p(2x2)surface], a change that is also calculated with the model
(18). The reason for this effect is that there are no closely
neighboring oxygen atoms on the p(2x2) surface, and the O_2
formation probability is much lower. Concepts of this sort may be
useful in testing for island-growth mechanisms and distinguishing
them from those that proceed through several distinct phases.

Energy Distributions

The energy distribution of atomic species ejected in bombardment
experiments are characterized by a peak at 1-5eV and a high energy
tail that goes approximately as E^{-n} where $n \approx 2$. This distribution
is characteristic of a non-equilibrium collision cascade. The
energy distributions of the parent molecular species are much
narrower, however, and often terminate at ~10eV. Shown in Figure
1a are experimental energy distributions for Ag^+, $C_6H_5^+$ and $C_2H_2^+$
ions ejected from a system with a monolayer of benzene adsorbed on
a Ag(111) crystal face (19). Since the molecular species is
ejecting during the same collision cascade as the Ag^+ ions and on
the same timescale one would expect the distribution of collision
energies that cause ejection to be the same for the Ag atoms and

the C_6H_6 molecules. However, the energetic collisions with the molecular species can and do cause fragmentation. Thus the energetic benzene molecules are depleted. The fragments then should have a distribution at higher energies as is illustrated by the $C_2H_2^+$ fragment energy distribution shown in Figure 1a. Note that the peak of the $C_2H_2^+$ distribution is at a higher energy than that of the $C_6H_5^+$ distribution. Since the peak position can be correlated to the binding energy of the species to the surface, the peak of the $C_2H_2^+$ distribution should be higher since its binding energy includes two C—C bond energies. The energy distributions from the calculations (Figure 1b) illustrate the same physical phenomena.

It is tempting to use the energy distributions of the ejected particles as a key to understanding the mechanisms responsible for the desorption. Care must be taken, however, as collision cascades can give rise to at least three distinctive shapes of energy distributions as shown in Figure 1. (The calculations also predict the distribution of metal atoms to have a high energy tail.) In fact the calculated C_6H_6 distribution of Figure 1b can be reasonably approximated by a Maxwell-Boltzmann form even though a thermal equilibrium is <u>not</u> present in the solid during the ejection event. The calculations indicate that energy distributions of elemental (and preferably both the neutral and charged) species could possibly be the most useful for comparing to the different experimental mechanisms as these particles cannot be fragmented in energetic collisions. Even here, however, one can obtain energy distributions from SIMS experiments that fall off more rapidly than E^{-2} if low energy ($\leqslant 250$ eV) primary ion beams are used (20).

Matrix Effects

The composition of the solid or matrix which is being bombarded has a large influence on the types of species observed to eject. This is true not only for the ionization process but also for the nuclear motion. Shown in Figure 2 are SIMS spectra of benzene taken for three different substrates. The data in Figure 2b was obtained for Ar^+ ion bombarded Ni(001) exposed to 3 langmuirs of benzene (16). This dose corresponds to approximately one monolayer coverage. This spectrum contains only the Ni^+, Ni_2^+ and $NiC_6H_6^+$ ions. Karwacki and Winograd also performed SIMS experiments for C_6H_6 adsorbed on Ni(001) where they dosed the surface with 2100 langmuirs of benzene (16). This SIMS spectrum is shown in Figure 2c. Here the multiple layers of benzene attenuate the Ni^+, Ni_2^+, and cationized $NiC_6H_6^+$ peaks. This spectrum, however, does contain hydrocarbon fragments of lower masses.

Two SIMS experiments have been performed on solid benzene at a temperature of 77 K (17,21). The mass spectrum from Lancaster et al. is shown in Figure 2d. They observe peaks at all masses corresponding to $C_nH_m^+$ where $n \leqslant 30$. The predominant peaks are the C_1, C_2, and C_3 species, in agreement with the work of Karwacki and Winograd (Figure 2c). We believe the reason we do not observe these $C_nH_m^+$ species with $n > 6$ in the calculations is due to the low density of benzene molecules on the Ni surface.

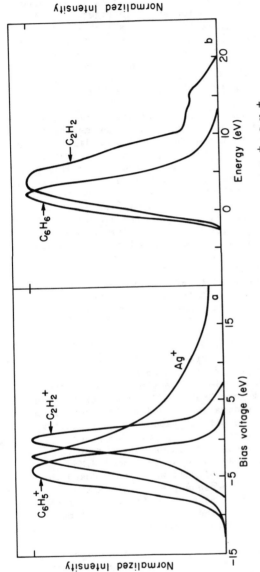

Figure. 1. Energy Distributions. a) Experimental Ag+, C6H5+
and C2H2+ ion intensities from one monolayer of benzene
adsorbed on Ag(111) plotted versus the voltage on the sample.
The primary particle is Ar+ with energy 1 keV incident on the
sample at a polar angle of 45°. The secondary particles are
collected at a polar angle of 60°. The raw data has been
smoothed. This data has been graciously provided by D. W.
Moon, R. J. Bleiler and N. Winograd prior to publication. b)
Calculated C6H6 and C2H2 energy distributions from one
monolayer of benzene on Ni(001). All ejected particles have
been counted. The energy resolution is 5eV. The calculations
are described in ref. 9. Reproduced with permission from Ref. 12.
Copyright 1983, Elsevier Science Publishing Co.

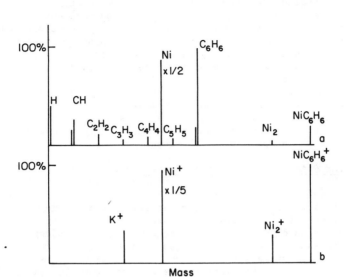

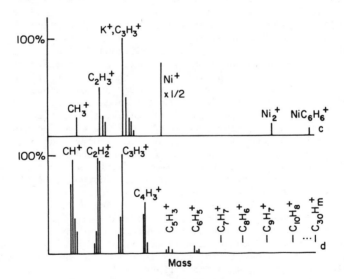

Figure 2. Benzene mass spectra. The most intense peak in each grouping has been identified. a) Calculated, (9). b) Experimental SIMS, 3 langmuirs of C_6H_6 on Ni(001), (16). c) Experimental SIMS, 2100 langmuirs of C_6H_6 on Ni(001), (16). d) Experimental SIMS, solid benzene (17). Reproduced with permission from Ref. 12. Copyright 1983, Elsevier Science Publishing Co.

It is obvious from Figure 2b-d that the sample preparation strongly affects the mass spectrum. The low coverage study appears to be the one where the parent species can be most easily identified as long as there is an energetically favorable means of ionization, e.g., cationization. The solid benzene studies are interesting in that a variety of large clusters are observed. However, if the sample were of an unknown compound, it would be difficult to extract the parent species from Figure 2d. The calculated spectrum (Figure 2a) predicts the parent molecule, C_6H_6, to be the most abundant organic species. The comparable experimental data, Figure 2b, however, has no $C_6H_6^+$ peak but a large $NiC_6H_6^+$ peak. Here then, the electronic environment influences which species are observed.

Molecular Orientation Effects: Benzene vs. Pyridine

It is of interest to compare the ejection mechanisms for molecules bonded to the surface with different orientations. In benzene, the interaction with the surface is shared among six carbon atoms via the π-electron cloud. In pyridine, however, the bonding occurs almost totally through the nitrogen atom while the remainder of the molecule is pointing away from the surface. The most striking difference between the two cases is that the computed yield of molecular species for the pyridine system is extremely low (9). The reasons for the major difference in yields for these two structures is clear from an analysis of the trajectories that lead to molecular ejection of pyridine. Very simply, pyridine ejection requires the specific cleavage of a N-Ni bond during a single collision. When a carbon atom is struck, the molecule either stays on the surface or tends to dissociate. There appears to be no efficient modes of transferring the energy of collisions with the molecule into translation away from the surface. Obviously the original structure of the organic molecules, then, affects the ejection and fragmentation processes. One would not necessarily expect similar spectra from a sample of a monolayer of organic molecules on a metal, a liquid, or an ordered solid.

These orientational effects have recently been confirmed in SIMS measurements of pyridine and benzene adsorbed on Ag(111) (22). In this system the benzene π-bonds to the surface while the pyridine π-bonds at low coverages but rearranges at higher coverages to σ-bond to the surface. The intensity of the $AgC_6H_6^+$ ion monotonically increases as the benzene coverage on the silver surface is increased to one monolayer. The $AgC_5H_5N^+$ and $C_5H_5NH^+$ ion intensities, however, initially increase and then decrease as the molecule rearranges on the surface, and finally increase again as the pyridine coverage is increased to one monolayer.

The arrangement of the molecules on the surface also influences the angular distributions of the ejected species (22). The polar angle distributions of various ejected ions for four systems −2.5 L benzene (monolayer), 4.5 L pyridine (monolayer, σ-bonded) 0.15 L pyridine (π-bonded), and 12 L thiophene (monolayer) on Ag(111) have been measured. The results of these distribution measurements are illustrated in Figure 3. For

monolayer benzene and for low coverage pyridine where the molecules
are believed to lie flat on Ag(111), the polar angle distribution
of $(M-H)^+$ (benzene) and $(M+H)^+$ (pyridine) are broad with a peak at
$\theta = 20°$ measured with respect to the surface normal. At the onset
of the change in bonding configuration, however, the polar angle
distribution of the $C_5H_5NH^+$ ion sharpens dramatically and the peak
moves to $\theta = 10°$. It appears that the array of σ-bonded pyridine
molecules provides a means of focusing the direction of ejection of
the pyridine molecules. Further, the polar angle distribution of
the high kinetic energy ions (6-10eV) ejected from the σ-bonded
pyridine structure is 20-30% wider than the distribution of the low
kinetic energy ions (3-7eV) as is shown in Figure 1b. This trend
toward wider polar angle distributions for faster moving particles
is counter to that observed for atom ejection. The polar angle
distribution of thiophene, is narrow with a peak at $\theta = 10°C$,
indicating that it also is σ-bonded to the surface.

In this case it appears that the σ-bonded pyridine molecules
are channeling the ejecting pyridine molecules into the vertical
direction (23). One example of how this blocking can significantly
affect the trajectory of an ejecting pyridine molecule is
illustrated in Figure 4. Only the species (one Ar^+ ion and two
pyridine molecules) directly involved in this particular ejection
process are shown. In this example the metal substrate plays no
direct role in ejecting the molecule. The grid lines are drawn
between the nearest-neighbor atoms in the first plane of the
microcrystallite. The elapsed time during the collision
process is shown in fs (1 fs=1x10^{-15} s). The initial positions of
the atoms are drawn in Figure 2 (0 fs). At 33 fs the Ar^+ ion,
which has backscattered from the surface, is colliding with 3
carbon atoms i the target pyridine molecule. The kinetic energy of
the center of mass of this pyridine molecule is 11.6 eV and its
molecular axis is oriented at a polar angle of $\theta=66°$ from the
surface normal. At 85 fs the ejecting pyridine molecule collides
with a neighboring pyridine molecule and dissipates a fraction of
its momentum. At the final stage of the sputtering process (120
fs), the pyridine molecule ejects molecularly, even though
distorted, at a polar angle of $\theta=31°$ with 1.40 eV of kinetic
energy. Due to the blocking by a neighboring pyridine molecule,
the polar angle of the ejected pyridine molecule is altered from
66° to 31°. The walls created by pyridine molecules are not
completely rigid as indicated by the distorted molecule shown on
the left in the 120 fs frame. Therefore, a pyridine molecule
ejecting with a large kinetic energy will not feel a strong enough
force to channel it completely into the upward direction. The
polar angle distribution of the high energy ejected particles is
thus broader than that of the low energy ejected particles. This
mechanism is distinct from that found with atom ejection. In this
latter case, the energy dependence of the azimuthal distribution is
related to the _time_ of ejection and consequently to the amount of
surface structure present when the atom ejects. Note that for the
π-bonded benzene system, there are no channels to orient the
ejecting molecules.

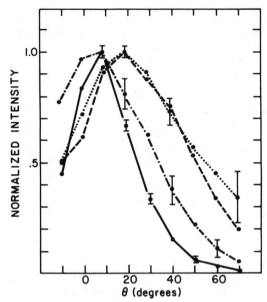

Figure 3. Normalized polar angle distribution of molecular ion yields for 4.5 L pyridine (———, $(M+H)^+$), 0.15 L pyridine (————, $(M+H)^+$), 2.5 L benzene (••••, $(M-H)^+$), and 12 L thiophene (—•—•—• —•—, M^+) on Ag(111) at 153K. The pyridine and benzene data is from (22) and thiophene data has been supplied by the same authors.

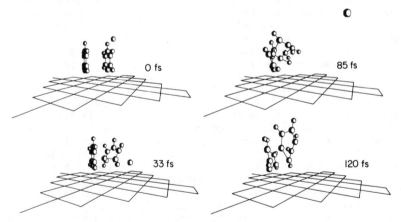

Figure 4. Change of the ejection angle of a sputtered pyridine molecule (right one) due to the blocking by a neighboring pyridine molecule (left one). The labels are in fs where 1 fs = 1×10^{-15} second. (0 fs) Initial positions of the atoms. (33 fs) The backscattered Ar^+ ion collides and ejects the pyridine molecule at a polar angle of $\theta = 66°$. (85 fs) The ejecting pyridine molecule is blocked by a neighboring pyridine molecule. (120 fs) Finally, the ejection polar angle is changed to $\theta = 31°$. Both the sputtered molecule and the blocking molecule are distorted.   Reproduced with permission from Ref. 23. Copyright 1985, Elsevier Science Publishing Co.

Fragmentation

There has been considerable speculation as to whether the observed
fragments form primarily from direct collisions at the surface
(i.e. within ~0.2 x 10^{-12}s after the primary particle has struck)
or from dissociation of larger species during the flight to the
detector (often as long as 10^{-4}s). The calculations show that it
is definitely possible to form numerous fragments in direct
collisions at the surface (Figure 2a). From the calculations we
have estimated that approximately three quarters of the ejected
benzene molecules have less than 5 eV of internal energy (9).
There is a reasonable probability that these vibrationally colder
molecules will remain intact. The energetic molecules, on the other
hand, will undoubtedly dissociate.

At this stage it is necessary to design clever experiments or
theoretical approaches to help elucidate the different possible
modes of fragmentation. Recently Moon (24) has proposed a method
of examining the polar angle distributions as a means of
differentiating between the fragmentation schemes. He finds that
for chlorobenzene adsorbed on Ag(111) that the $C_6H_5^+$ and Cl^-ions
probably form by direct collisions on the surface. For the
chlorobenzene as well as benzene and pyridine adsorbed on silver,
he found that neither molecular or fragment ions formed by gas
phase decomposition of a cationized species.

In the case of the alkali halide clusters (14), recent work
has shown that the oscillations in ion intensity with cluster size
are due to dissociation of metastable species during the flight to
the detector (25). Spectra taken 0.2µs after bombardment exhibit a
monotonic decrease in ion intensity with increasing cluster size.
Spectra taken after 70µs, however, show an increase in the
$(CsI)_{13}Cs^+$ ion intensity and a decrease in the $(CsI)_{14}Cs^+$ and
$(CsI)_{15}Cs^+$ ion intensities. Here then, decomposition of larger
species during the flight to the detector has a noticeable effect
on the cluster yields. These experiments though make no statement
as to how the clusters are initially formed near the surface.

Closing Statements

A classical dynamics model has been developed to investigate the
importance of collisional processes in heavy particle bombardment
experiments. This procedure is very powerful for describing
collisional events and provides a working hypothesis against which
experimental data can be compared. We have shown numerous examples
from SIMS experiments where the calculations have fit experimental
data very well. The time has come for the experimentalists to
conceive and execute experiments aimed at uncovering the
fundamental processes involved in the SIMS and FABMS procedures.

It should be noted that various researchers have different
goals for using and understanding the ion bombardment process.
There are those who are using the technique to obtain information
about a molecule that they have placed on the surface. That is,
they want a mass and possibly a structure determination. Other
researchers are primarily concerned with determining the elemental
composition of the sample while others use the technique to measure

the geometrical arrangement of the atoms and molecules on the surface. Another area of interest is to probe the electronic processes involved when an atom or molecule is in the near surface region. Although these goals are quite varied the fundamental processes are intermingled. To understand our own area of interest we need to understand all of the experimental results and the detailed events occurring on the microscopic level.

Acknowledgment

The interaction with those who have supplied the experimental data, D. W. Moon, R. J. Bleiler, E. J. Karwacki and N. Winograd, has greatly helped in solidifying many of the ideas presented here. I thank them for allowing me to use their data as well as for many stimulating conversations. The financial support of the National Science Foundation, the Office of Naval Research and the Camille and Henry Dreyfus Foundation is gratefully acknowledged.

Literature Cited

1. McNeal, C. J.; Macfarlane, R. D. J. Am. Chem. Soc. 1981, 103, 1609.
2. Orth, R. G.; Jonkman, H. T.; Michl, J. J. Am. Chem. Soc. 1981, 103, 1564.
3. Garrison, B. J.; Winograd, N.; Harrison Jr., D. E. J. Chem. Phys. 1978, 69, 1440.
4. Winograd, N.; Harrison Jr., D. E.; Garrison, B. J. Surface Science 1978, 78, 467.
5. Garrison, B. J.; Winograd, N.; Harrison Jr., D. E. Phys. Rev. B 1978, 18 6000.
6. Winograd, N.; Garrison, B. J.; Harrison Jr., D. E. J. Chem. Phys. 1980, 73, 3473.
7. Foley, K. E.; Garrison, B. J. J. Chem. Phys. 1980, 72, 1018.
8. Garrison, B. J. J. Am. Chem. Soc. 1980, 102, 6553.
9. Garrison, B. J. J. Am. Chem. Soc. 1982, 104, 6211.
10. Garrison, B. J. J. Am. Chem. Soc. 1983, 105, 373.
11. Winograd, N.; Garrison, B. J. Accts. of Chem. Res. 1980, 13, 406.
12. Garrison, B. J.; Winograd, N. Science 1982, 216, 805; Garrison, B. J. Int. J. Mass Spec. and Ion Phys. 1983, 53, 243.
13. Garrison, B. J.; Winograd, N. Chem. Phys. Lett. 1983, 97, 381.
14. Barlak, T. M.; Wyatt, J. R.; Colton, R. J.; Decorpo, J. J.; Campana, J. E.; Campana, J. E.. J. Am. Chem. Soc. 1982, 104, 1212.
15. Brenner, D. W.; Garrison, B. J. unpublished results.
16. Karwacki, E. J.; Winograd, N. Anal. Chem. 1983, 55, 79.
17. Lancaster, G. M.; Honda, F.; Fukuda, Y.; Rabalais, J. W. J. Am. Chem. Soc. 1979, 101, 1951.
18. Winograd, N.; Garrison, B. J.; Fleisch, T.; Delgass, W. N.; Harrison Jr., D. E. J. Vac. Sci. Tech. 1979, 16, 629.
19. Moon, D. W.; Bleiler, R. J.; Winograd, N. unpublished results.

20. Hart, R. G.; Cooper, C. B. Surf. Sci. 1979, 82, 549; Karwacki,
    E. J. Ph.D. thesis, The Pennsylvania State University, 1982.
21. Jonkman, H. T.; Michl, J.; King, R. N.; Andrade, J. D.
    Anal. Chem. 1978, 50 2078.
22. Moon, D. W.; Bleiler, R. J.; Karwacki, E. J.; N. Winograd.
    J. Am. Chem. Soc. 1983, 105, 2916.
23. Moon, D. W.; Winograd, N.; Garrison, B. J. Chem. Phys. Lett.,
    in press.
24. Moon, D. W.; Winograd, N. Int. J. Mass Spec. and Ion Phys.
    1983, 52, 217.
25. Ens, W.; Beavis. R.; Standing, K. G. Phys. Rev. Lett. 1983, 50,
    27.

RECEIVED August 23, 1985

3

Role of Intermolecular Interactions in the Desorption of Molecular Ions from Surfaces

Ronald D. Macfarlane

Department of Chemistry, Texas A&M University, College Station, TX 77843

Intermolecular interactions of surface molecules can
modify the spectrum of molecular ions that are
desorbed when the surface is excited. These
interactions may be important in all of the particle
induced desorption processes including SIMS and FAB
and may involve the manner in which excitation energy
from the primary ion is dissipated into the medium.
A brief account of some experiments being carried out
to gain a better understanding of the energy
relaxation process for high and low energy ion
bombardment is given. Experiments using Langmuir
adsorption to control intermolecular spacing between
Rhodamine molecules are described with a
demonstration of the influence on the mass spectra of
these species. Finally, some experiments on the
desorption of molecular ions of insulin will be
discussed.

This symposium is devoted to a discussion of the similarities and
differences of SIMS and FAB, the two most popular particle-induced
desorption methods. There is another method called 252-Californium
plasma desorption mass spectrometry (252-Cf PDMS) which was
introduced in 1974 that bears a strong resemblance to these methods
primarily from the similarities of the main features of the mass
spectrum of the desorbed molecular ions (1). It is probably for this
reason that 252-Cf PDMS was included in this Symposium because it may
be part of the complex picture of how these methods work. That the
mass spectra are remarkably similar in most cases, particularily for
complex biomolecules, is an important observation because it implies
that the particular molecular excitations that are operating in the
emission-ionization process may be insensitive to the details of the
primary energy form brought in by the incoming ion or atom, that it
does not matter whether the primary particle is an ion, atom, photon,
or is at keV or 100 MeV energy. This remarkable universality is one
of the puzzling aspects of the problem. Models have been introduced
to give some physical picture to what might be going on and while

0097–6156/85/0291–0056$06.00/0
© 1985 American Chemical Society

these have given conceptual insights they may also have contributed to the present state of confusion. The problem is that there are not enough experimental facts on which to base a sound model. However, there are some data, some from fields not associated with mass spectrometry, that may be giving some clues to what is going on. The first part of this paper includes some of these results. The second part describes some of our recent studies that relate to the role of intermolecular interactions in mediating the emission-ionization process. Finally, results on the desorption of massive molecular ions of biopolymers will be discussed; there seem to be some real differences between low and high energy particle-induced emission for these species.

The Energy Deposition Process

It is now clear that it does not matter whether a primary particle is neutral or charged and that the fundamental differences between SIMS and FAB lie within the use of a solid vs. liquid matrix. In subsequent discussions in this paper no distinction will be made between neutral and ionic primary particles. The SIMS and FAB method utilize primary ions in the 5-20 keV regime where nuclear stopping is known to be the primary mechanism for energy deposition. The initial impact triggers a collision cascade which results in the sputtering of secondary atomic ions with a relatively high kinetic energy distribution, a signature of processes supported by an energetic collision cascade. This phase of the interaction is well supported by experimental data and a quantitative theory, the Sigmund sputtering theory (2). Molecular ions emitted in the process have a much lower kinetic energy distribution which cannot be fitted to Sigmund theory suggesting that they are not emitted during the part of the collision cascade that produces atomic ions (3). Recently, Standing studied the emission of molecular ions of simple amino acids as a function of primary ion energy using ions ranging from Li to Cs and energies from 1 to 16 keV (4). There were some important conclusions in this study: molecular ion yields closely follow the amount of energy brought in by the nuclear stopping mechanism and for Li primary ions where nuclear and electronic stopping both contribute to the energy deposition it was the total energy deposited that was the significant factor in determining the molecular ion yield. Electronic excitation is different from nuclear excitation because the excitation involves promoting electrons in the matrix to higher excitation states whereas nuclear excitation moves atoms from their positions in the matrix. The Standing results give experimental evidence that despite these large differences in the initial form of excitation of the matrix, the emission of secondary molecular ions is insensitive to these differences.

A considerable amount is known about details of the primary events in electronic excitation. Electronic excitation cross sections are dependent on the charge and velocity of the incident primary ion. Both of these parameters have been confirmed to be the important variables in the emission of secondary molecular ions under bombardment by ions in the 50-100 MeV regime (1 MeV/u) (5-6). The excitation promotes electrons within atoms and molecules to very high energies resulting in ionization and emission of secondary electrons (up to 50 per incident ion) (7). The excitation of electrons in core

levels produces vacancies that are refilled by Auger processes
resulting in atomic ions with multiple charge states. These are
part of the spectrum of atomic ions emitted in 1 Mev/u ion
bombardment and are totally absent in SIMS spectra using the same
targets. Weller has recently made this comparison and has extended
the energy range for SIMS primary ions to 100 keV (8).

High energy ions entering a solid produce a linear damage track
of about 15 microns in length that can be visualized by electron
microscopy (9). The linear energy transfer (LET) to the matrix is a
measure of the density of energy deposition which is a function of
the charge and velocity of the primary ion. By varying the charge of
the primary ion and with a knowledge of the charge equilibration
process, Voit has shown that only the energy deposited within the top
6-10 molecular layers of a valine matrix is contributing to the
emission of secondary molecular ions. This means that for a 100 MeV
fission fragment ion, as is used in 252-Cf-PDMS, only 50 keV is used
in the emission of secondary ions (LET=10,000 ev/nm) but this is
concentrated in a very small volume at the surface, and in a
subfemtosecond time frame, giving an effective power density of 10^{15}
watts/cm^2. The profile of a heavy ion track contains an inner core
with a radius of 0.1 nm that includes all the atoms and molecules
that were excited by the primary ion. This is surrounded by a
cylinder of energetic electrons (delta rays) extending to 1 nm that
were ejected from the inner core in the initial excitation process
(10). The lifetime of this track structure is approximately 10^{-17}s.
There is not much experimental information as to what happens after
this.

The Energy Dispersion Process

While the primary excitation processes for both keV and MeV ions is
well understood, little is known on how this energy is transformed
into the form required for the ejection of secondary molecular ions.
In both cases, the volume excited, if crystalline, is melted and
freezes to an amorphous state (11-12). This is an experimental
observation indicating that at some stage in the energy dispersion
lattice vibrations are excited to a level where intermolecular bonds
are broken. We have some experimental clues as to what happens after
the primary track is formed by a high energy ion. If the track is
formed in a semiconductor, it is possible to collect and measure the
electron current produced in the generation of electron-hole pairs.
This is the basis for the operation of semiconductor detectors used
in nuclear spectroscopy. For ions with a high LET, the density of
electron-hole pair formation is so great that recombination processes
occur (non-linear processes) resulting in a loss of electron current
(ionization defect) (13). When the matrix is an insulator, energy
transfer is mediated by various forms of excitons which can propagate
through the solid. The exciton concept was first introduced by
Forster as a mechanism for energy transfer in solids. The existence
of excitons (quanta of electronic excitation) is now well documented
and its role in secondary ion emission as now been experimentally
verified for argon atoms sputtered by MeV He ions from argon ice
(15). Exciton migration to fluorescent sites in plastic
scintillators is the mechanism of the operation of plastic
scintillator crystals as particle detectors for nuclear spectroscopy.

multiphonon absorption into the bonds holding the molecule to the surface (22). The key measurements to be made that would contribute to an understanding of this phase of the problem are the kinetic energy and angular distribution of the ejected ions and molecules. Winograd has made these measurements for SIMS using well-defined adsorbate-substrate systems and Wien has made similar measurements for 252-CF-PDMS (23,24).

Perhaps the most confusing aspect of the problem is the ionization mechanism. It is difficult to control the matrix to the level where significant progress can be made in understanding what processes are involved that produce the ionized molecule. The simplest species to study is the preformed ion, a species in the condensed phase that has lost or gained most of an electron as a consequence of electronegativity differences. Whether it remains an ion in the emission process depends on the electronic interactions operating at the surface between adsorbate and substrate (25). The larger the ion, the greater is the probability that the preformed state is retained. Large organic cations (quaternary ammonium salts) and large organic anions (oligonucleotides) are examples. Small anions such as Cl⁻ in NaCl are apparently emitted as neutral species while multiply charged species are emitted as singly charged ions (26,27). These observations suggest that surface processes can play a role in determining the ultimate charge state of a desorbed species.

When one considers the role of the matrix in the particle-induced emission of secondary ions it is no wonder that it is so difficult to unravel all the processes that take place. The matrix is the medium in which the primary excitation occurs. It must also disperse some of that energy to sites at the surface where secondary ion emission occurs. It must provide the species to be desorbed and at the same time mediate the ionization process. In an attempt to understand these complex coupled processes we have tried to simplify the system by first selecting a homogeneous substrate for the energy deposition and then studying the ionization-emission process for species that are present as a submonolayer on the surface (28).

Polymeric Surfaces as Substrates

In a 252-Cf-PDMS measurement, the sample consists of a thin metallized backing on which the sample is deposited using techniques such as the electrospray method (29). The conducting surface is required in the measurement because it establishes the electric field lines through which the ejected ions are accelerated. We have recently found that if we reverse the metallized polymer film so that the polymer side is the desorption surface, it is still possible to obtain good quality mass spectra that show no evidence for electrostatic charging. This geometry is depicted in Figure 1. The significance of this is that the polymer surface can be utilized to selectively adsorb solute molecules from solutions in a manner analogous to the adsorption of solutes onto the stationary phase in liquid chromatography columns. The number density of solute molecules on the surface can be controlled by varying the solution concentration and solvent type. In the first experiments the native polymer surface of Mylar (polyethylene terephthalate) was employed and solute molecules were adsorbed onto the Mylar surface from

When these detectors are used for the high LET fission fragments the
light output is considerably reduced relative to other ions with
comparable energy but lower LET (16). This has been attributed to
the quenching of exciton states in the heavy ion track due to their
high density. These radiationless transitions deposit their energies
into vibrational excitation within the matrix. The exciton model is
well developed. There are many different types of excitons (e.g.
atomic and molecular). Kimura has shown that alkali halides excited
by high energy ions first form atomic excitons that localize on
individual ions within the crystal lattice but these ultimately decay
into molecular excitons that are associated with diatomic states and
which fluoresce with a characteristic spectrum (17).

An attractive feature of the exciton model is entry into the
family of exciton states can be made by either electronic excitation
(transverse excitons) as with high energy ions or by collisional
processes (longitudinal excitons) and both could decay into the same
pattern of electronic and vibrational excitations (18). This could
provide a microscopic explanation for the similarities of SIMS and
252-Cf-PDMS mass spectra. If the emission of secondary ions from
surfaces is mediated by exciton-phonon processes, the energy
dispersion will be influenced by physical features of the matrix
(degree of crystallinity and purity) which are not easily controlled.
For species with a multiplicity of intermolecular interactions and
which do not exist as ions in the solid state, the probability for
secondary molecular ion formation is dependent on properties of the
matrix which are not understood (19). The relationship of molecules
at the surface which are participating in the formation of secondary
ions with neighboring molecules may also be important in the final
stages of the energy dispersion process. Some experiments directed
to this question will be described later in this paper.

The Emission-Ionization Process

The fact that secondary ions are indeed observed in SIMS (FAB) and
252-Cf-PDMS means that some of the energy dispersion appears in the
form needed to effect emission of molecular species from the surface
which ultimately appear as ions in a mass spectrometer. Standing has
made the point that instrumental parameters may be influencing the
ion formation step (20). The time-of-flight mass spectrometers used
in 252-Cf-PDMS and in some SIMS studies are primarily sensitive to
ion-forming processes that occur at or very close (10 nm) to the
surface while in most magnetic sector and quadrupole instruments, the
ionization region may be considerably extended to the point where gas
phase ion-molecule reactions might occur. This ambiguity may be the
source of some of the disagreements with regard to where the
ionization is occurring. Gas phase ion-molecule reactions such as
occur in chemical ionization are better understood than the processes
that occur at surfaces. The mechanism of the emission process has
been the subject of several theoretical treatments. One of these
considers that the emission process is associated with the later
stages of the collision cascade and that even in the case of an
initial primary electronic excitation a collision-like cascade is
developed (21). Another approach considers the energy to be so well
dispersed that the emission site is vibrationally excited. The
molecule is ejected from the surface as a consequence of a

solutions of polar solvents (water, methanol, ethanol) utilizing the
hydrophobic interaction (30).   In subsequent experiments the Mylar
surface was chemically modified by coating with polymers having
different functional groups (cation exchange, anion exchange) so that
the adsorption properties could be varied over a wide range.   This
arrangement of substrate-adsorbate made it possible to utilize a
fixed homogeneous matrix where the primary energy deposition and
dispersion occurred.   In addition, by varying the concentration of
the solution, it was possible to produce films with different number
densities of solute species.   The questions to be answered initially
were: does the nature of the substrate influence the mass spectrum
of ions that are emitted from the surface? Does the surface
concentration of solute molecules influence the mass spectrum of
secondary ions in ways other than effecting the overall intensity of
the ions?

Rhodamine 6-G.   This molecule was selected for the first study
because its adsorption characteristics on surfaces had already been
studied.   The fluorescence characteristics of this molecule when
adsorbed on a surface provides an internal measure for the average
distance between molecules (31).   In addition the solid exists in the
form of a hydrochloride salt and the cation exists essentially as a
large preformed ion.   The structure of the molecule has a large
aromatic component so that adsorption on the Mylar which also has an
aromatic ring in its structure is favorable.   The 252-Cf-PDMS
spectrum of an electrosprayed deposit of this molecule is dominated
by an intense molecular ion with very little fragmentation.   The
study involved the preparation of a series of ethanol solutions of
Rhodamine 6-G with concentrations varying from 10^{-5} M to 10^{-2} M.   A
100ul aliquot was deposited on the surface of the Mylar and
equilibrated for a few minutes.   The droplet was then removed and the
sample was inserted into the mass spectrometer for mass analysis.
This procedure was repeated using the different solution
concentrations giving a series of targets with increasing
concentrations of the solute on the Mylar surface.   Figure 2 shows
the 252-Cf-PDMS spectrum of one of the Mylar films containing
Rhodamine 6-G adsorbed from a 10^{-5} M solution.   Figure 3 is a plot of
the intensity of the Rhodamine 6-G molecular ion as a function of
solution concentration.   The reproducibility of the measurement is
10%.   An independent measurement of the number density of Rhodamine
6-G by a Beer-Lambert absorbance confirmed the results of earlier
measurements of Garoff using optical spectroscopic techniques (31).
The intensity-concentration curve is characteristic of a Langmuir
adsorption mechanism.   It was possible to deduce from the linear
portion of the curve that the effective region of excitation for ion
emission is relatively small, much smaller than the 20 nm diameter
damage areas measured by electron microscopy for equivalent samples
(12).   If the excitation area were large, at a particular number
density the area would have intercepted more than one molecule.   This
would have resulted in a positive deviation in the linear portion of
the adsorption curve and none was observed.   Two effects were
observed that relate to the influence of the number density at the
surface on the nature of the mass spectrum of secondary ions.   First,
the Cl counter ions were only observed when monolayer coverage was
realized.   This implies that Cl is emitted as a neutral species at

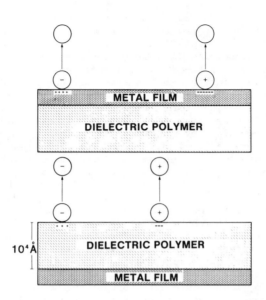

Figure 1:   Schematic representation of the aluminized Mylar
surface.   The polymer side of the film is used for the
adsorption studies.

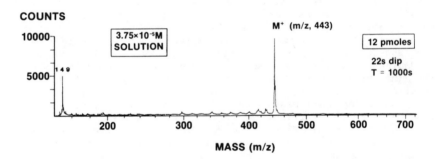

Figure 2:   252-Cf-PDMS spectrum of Rhodamine 6-G adsorbed from
an ethanol solution onto Mylar.

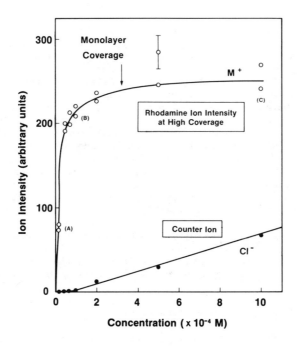

Figure 3:   The   molecular   ion   intensity   of   Rhodamine · 6-G
(determined   by   252-Cf-PDMS)   as   a   function   of   solution
concentration.

low coverage but as an anion at high coverage.   While there was no
detectible change in the mass spectrum in the region of the molecular
ion, there were significant differences in the dimer ion region.
This is shown in Figure 4.   At low coverage there is no evidence for
dimer ion formation.   At intermediate coverage, a broad distribution
appears which is characteristic of metastable ion emission.   This we
attribute to the formation of contact pair dimers which are randomly
aligned in the matrix. At high coverage where the solution contained
a significant component of stable dimers there was a significant
intensity of dimer-related ions in the mass spectrum.   This first
study confirmed the feasibility of carrying out mass measurements for
a series of molecules with varying intermolecular separations.

Rhodamine-B.   This molecule is closely related to Rhodamine 6-G in
structure as well as in its optical properties.   The 252-Cf-PDMS
spectrum, however does have an important difference.   It contains a
fragment ion corresponding to loss of a carboxyl group.   Since
fragmentation is a measure of vibrational excitation within the
molecule, it was possible to determine whether two Rhodamine-B
molecules in close proximity behaved any differently from molecules
isolated from each other on the surface.   Figure 5 shows the results
of this experiment.   At low surface concentrations, the fragment and
molecular ion intensity are comparable while at high coverage the
fragment ion intensity is noticeably diminished.   This observation
suggests that the excitation energy can be shared amongst molecules
when they are in close proximity.   In a very recent measurement where
the Rhodamine-B was adsorbed on an evaporated Au layer, the fragment
ion was essentially absent.   This indicates that the degree of
molecular excitation can also be controlled by changing the
substrate.

Emission of Massive Molecular Ions From Surfaces

Part of the current interest in the use of particle-induced emission
for the mass spectrometry of organic species is that it has been
possible to desorb ions of very large complex biomolecules like
insulin.   The current record is for the protein trypsin which has a
molecular weight of 23,000 u (32).   It is in the study of these
species by the various particle-induced desorption methods that
differences are beginning to appear between FAB, SIMS, and 252-Cf-
PDMS.   For the keV primary ion work (SIMS, FAB) differences are
appearing that seem to be associated with the matrix.   FAB has been
more successful in producing these ions than SIMS.   The reasons are
not yet understood.   It has been suggested by several investigators
that the difference is associated with the higher mobility of solute
species in the liquid matrix, but perhaps the liquid matrix is
supporting the generation of an aerosol in the ion source or is
providing a region of high local pressure where gas phase processes
might occur.   Benninghoven, earlier in the studies of molecular SIMS
introduced the concept of a damage cross section to explain the
observation that the yield of secondary ions decreases when the
fraction of the surface not perturbed by the effects of the collision
cascade diminishes (33).   The suggestion was that the damaged area is
actually depleted in molecules that could contribute to the
desorption yield.   The "damage cross section" is precisely a measure

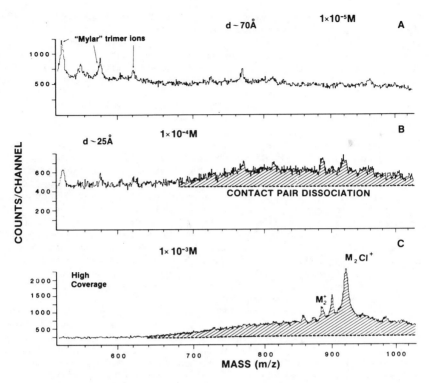

Figure 4:   252-Cf-PDMS spectrum of Rhodamine 6-G in the dimer region for various solution concentrations.

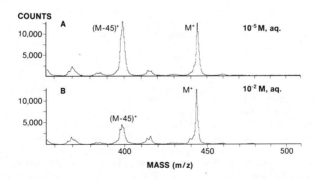

Figure 5:   252-Cf-PDMS spectrum of Rhodamine-B for films prepared from two different solution concentrations.

of the loss of the ability of the matrix to desorb species as ions
and there may be several factors that could contribute to this
effect.   This is also observed for high energy ion excitation (34).
The SIMS studies referred to earlier suggests that the damage regions
represent parts of the matrix that have become amorphous (11).   This
means that when the molecules in the matrix become randomly oriented
that part of the matrix cannot support the emission of molecular
ions.   For FAB however, the molecules in the excitation region only
temporarily lose their relative alignment.   Their high mobility in
the fluid matrix gives them an opportunity to restore their preferred
orientation which presumably enhances their intermolecular
interactions.   The implication here is that the formation of
molecular ions may be mediated by the intermolecular interaction.
        At the present, the FAB method has been able to produce
molecular ions of biomolecules up to m/z 10,000 (35).   Molecular ion
yields are decreasing orders of magnitude with each 1000 u increase
in mass.   This is not the case for 252-Cf-PDMS.   Other differences
are beginning to show up.   The average number of secondary ions
emitted per incident ion is a factor of 4 higher for high energy
primaries and multiplicities as high as 36 have been detected for
some fission tracks (36,37).   Sundqvist has shown that the desorption
of large molecular ions requires the high energy density of the track
produced by a high energy primary ion.   The yield of insulin
molecular ions varies as the LET to the sixth power while for small
molecules the yield varies as the LET to the first or second power.
This means that non-linear processes are involved in the desorption
of these species.   What these are is not known but it may be that the
quenching processes observed in heavy ion tracks due to high
concentrations of electron-hole pairs or molecular excitons may be
contributing to the desorption of large, complex molecular ions.   The
LET of a typical SIMS primary ion is a factor of 10 lower than for a
252-Cf fission fragment.   This translates to a six orders of
magnitude enhancement for the yield of insulin for 252-Cf-PDMS
relative to SIMS using a solid matrix.   Processes occurring in the
liquid matrix of FAB may enhance this yield.
        The role of the matrix in the formation of molecular ions of
large complex biomolecules seems to be more acute than with smaller
molecules.   It is hoped that the studies using adsorbed species on
different surfaces will give insights to the optimum matrix for the
desorption of these larger ions.

Conclusions

It is the intent of this Symposium to make some progress in the
understanding of the similarities and differences of SIMS and FAB
when these methods are used to produce molecular ions of
biomolecules.   Certainly some progress has been made in understanding
some of the similarities in terms of the mode of primary energy
transfer.   Liquid matrix vs. solid matrix is a much more difficult
question but the experimental facts at this time support the notion
that the liquid matrix, under high primary flux excitation, has
features that are not present in a solid state SIMS experiment.
Where does 252-Cf-PDMS fit in this picture?   Is it really SIMS and
nothing more?   Perhaps this is a question more philosophical than
substantive at this point.   The 252-Cf-PDMS method is a maverick in

the mass spectrometry community. None of its features fit into the mold of acceptable mass spectrometric methods but this is for us an attraction rather than deterrent. The historical perspective of mass spectrometry has a significant component of Darwinian evolution....the survival of the fittest ....of methods. The survival of 252-CF-PDMS, or SIMS or FAB as a species for mass spectrometry of biomolecules will be dependent on their relative utility over a long time span; in 252-Cf-PDMS, time is a comfortable variable.

Acknowledgments

It is a pleasure to acknowledge the many contributions of my colleagues at Texas A&M to the work described here, particularily Dr. C.J. McNeal, V. Mancell, and E.A. Jordan. The support of the National Institutes of Health (GM-26096), the National Science Foundation (CHE-82-06030) and the Robert A. Welch Foundation (A-258) made this work possible.

Literature Cited

1. Torgerson, D.F.; Skowronski, R.P.; Macfarlane, R.D. Biochem. Biophys. Res. Commun. 1974, 60, 616.

2. Sigmund, P. J. Vac. Sci. Technol. (N.Y.) 1980, 17, 396.

3. Smith, R.D.; Burger, J.E.; Johnson, A.L. Anal. Chem. 1981, 53, 1603.

4. Standing, K.G.; Chait, B.T.; Ens, W.; McIntosh, G.; Beavis, R. Nucl.Instrum. Methods 1982, 198, 33.

5. Hakansson, P.; Jayasinghe, E.; Johansson, A.; Kamensky, I.; Sundqvist, B. Phys. Rev. Lett. 1981, 47, 1227.

6. Hakansson, P.; Sundqvist, B. Radiat. Eff. 1982, 61, 179.

7. Della Negra, S.; Jacquet, D.; Lorthiois, I.; Le Beyec, Y. Int. J. Mass Spectrom. Ion Phys. 1983, 53, 215.

8. O'Connor, J.P.; Blauner, P.G.; Weller, R.A. Yale University Report (Yale-3074-804), 1985.

9. Fleischer, R.L.; Price, D.B.; Walker, R.M. "Nuclear Tracks in Solids"; University of California Press: Berkeley, 1975.

10. Ritchie, R.H.; Claussen, C. Nucl. Instrum. Methods 1982, 198, 133.

11. Fujimoto, J.G.; Liu, J.M.; Ippen, E.P.; Bloembergen, N. Phys. Rev. Lett. 1984, 53, 1837.

12. Bowden, F.P.; Chadderton, L.T. Proceed. Royal Soc.,A, 1962,269,143.

13. Finch, E.C.; Cafolla, A.A.; Asghar, M. Nucl. Instrum. Methods, 1982, 198, 547.

14. Forster, Th. In Modern Quantum Chemistry, Vol. III; Sinanoglu, Ed.; Academic Press: New York, 1965, 93.

15. Reimann, C.T.; Johnson, R.E.; Brown, W.L. Phys. Rev. Lett., 1984, 53, 600.

16. Neiler, J.H.; Bell, P.R. In Alpha-, Beta-, and Gamma Ray Spectroscopy, Vol. I; Siegbahn, K. Ed.; North-Holland, Amsterdam, 1965, 245.

17. Kimura, K.; Mochizuki, K.; Fujisawa, T.; Imamura, M. Phys. Rev. Lett. 1980, 78A, 108.

18. Kasha, M. In Spectroscopy of the Excited State; DiBartolo, B., Ed.; Plenum: New York, 1976; p 337.

19. Macfarlane, R.D. Acc. Chem. Res. 1982, 15, 268.

20. Ens, W.; Beavis, R.; Standing, K.G. Phys. Rev. Lett. 1983, 50, 27.

21. Garrison, B.J. Int. J. Mass Spectrom. Ion Phys. 1983, 53, 243.

22. Kreuzer, H.J. Int. J. Mass Spectrom. Ion Phys. 1983, 53, 273.

23. Winograd, N. Proceedings of this Conference.

24. Furstenau, N.; Knippenberg, W.; Krueger, F.R.; Weiz, G.; Wien, K. Z. Naturforsch. A, 1977, 32A, 711.

25. Sroubek, Z. Int. J. Mass Spectrom. Ion Phys. 1983, 53, 289.

26. Cerny, R.L.; Sullivan, B.P.; Bursey, M.; Meyer, T.J. Anal. Chem. 1983, 55, 1954.

27. Macfarlane, R.D. (unpublished results)

28. Macfarlane, R.D. Proc. Conf. Mass Spectrom Life Sciences, Univ. of California, San Francisco, 1984 (in press).

29. McNeal, C.J.; Macfarlane, R.D.; Thurston, E.L. Anal Chem. 1979, 51, 2036.

30. Prieto, N.E.; Martin, C.R. J. Electrochem. Soc. 1984, 131, 751.

31. Garoff, S.; Stephens, C.; Hanson, C.D.; Sorenson, G.K. Optics Commun. 1982, 29, 57.

32. Sundqvist, B; Roepstorff, P.; Fohlman, J.; Hedin, A.; Hakansson, P.; Kamensky, I.; Lindberg, M.; Salehpour, M.; Sawe, G. Science, 1984, 226, 696.

33. Benninghoven, A. Int. J. Mass Spectrom. Ion Phys., 1983, 46, 459.

34. Salehpour, M.; Hakansson, P.; Sundqvist, B. Nucl. Instrum. Methods, 1984, 82, 752.

35. Barber, M.; Bordoli, R.S.; Elliot, G.J.; Horoch, N.J.; Green, B.N. Biochem. Biophys. Res. Commun. 1983, 110, 753.

36. Ens, W. Ph.D. Thesis, University of Manitoba, Winnipeg, 1984.

37. Sundqvist, B.; Hakansson, P.; Kamensky, I.; Kjellberg, J. Int. J. Mass Spectrom. Ion Processes 1984, 56, 52.

38. Sundqvist, B.; Hedin, A.; Hakansson, P.; Kamensky, I.; Kjellberg, J.; Salephour, M.; Sawe, G.; Widdiyasekera, S. Int. J. Mass Spectrom. Ion Phys. 1983, 53, 167.

RECEIVED April 16, 1985

Processes of Laser-Induced Ion Formation in Mass Spectrometry

F. Hillenkamp[1], M. Karas[1], and J. Rosmarinowsky[2]

[1] Institute of Biophysics, University of Frankfurt, Abtlg. für Biophysikalische Strahlenforschung, Frankfurt, Federal Republic of Germany
[2] Gesellschaft für Strahlen- und Umweltforschung, München, Federal Republic of Germany

In its first part the paper discusses properties of laser desorption spectra of nonvolatile, fragile organic compounds as well as current models, suggested for the relevant ion formation processes. In the second part some recent results on metastable decay and the influence of the laser wavelength on desorption spectra are presented.

The generation of ions from solid samples by high power laser pulses has been intensely investigated in fusion research for over 20 years. The hot plasmas, produced in these experiments contain predominantly multiply charged ions of high kinetic energy (33). The generation of ions out of the condensed phase with lasers for analytical purposes in mass spectrometry has only more recently gained considerable interest. Three fields of application justify this interest:

1. Laser beams can be focused down to the diffraction limit of 0.2 - 0.5 μm with very high efficiency. This has lead to the development of the Laser-Mass-Spectrometric Microprobes LAMMA 500^R, LAMMA 1000^R and LIMA. These instruments are designed for tasks in principle comparable to those of dynamic SIMS or imaging ion-microprobes.

2. A convincing body of experimental information, now available in the literature, serves as confirmation for the possibility to desorb molecular ions out of the condensed phase even for organic molecules which are generally considered nonvolatile and/or fragile and do therefore not lend themselves to classical mass spectrometric analysis. Here the laser-MS competes with techniques such as static SIMS or FABMS, plasma- and field-desorption.

3. At suitably high laser irradiances at the solid surface, a plasma is formed as mentioned above. With properly chosen irradiation parameters, this plasma shows features, comparable to those of HF-spark sources, commonly used in mass spectrometry of inorganic solid samples. From results, reported in the literature (1, 2, 3), it can be deduced that the laser source may offer some advantages over classical spark sources with respect to sample preparation (nonconductors), reproducibility, collection efficiency and uniformity of elemental sensitivity factors. The laser in addition

0097-6156/85/0291-0069$06.00/0

offers the advantage of high spatial resolution as described
in 1.

The results reported by the different groups have all been
obtained with experimental arrangements that differ greatly in the
types of laser used, the wavelengths, time regime- and irradiances
on the sample, in the sample geometry and sample preparation, the
mass spectrometers and the detection systems. The most pertinent information on the different systems that have been successfully applied
to organic mass spectrometry is compiled in ref. (4).

This paper contains a discussion of the most important features
of laser-induced-ion spectra and mechanisms of ion formation, some of
them experimentally proven, some of them still under discussion and
investigation. Special attention will be given to the LAMMA technique,
because this is the authors own field of work, but reference to other
systems will be included where appropriate.

Characteristics of spectra

Ion spectra of inorganic as well as of organic samples have been
published by many authors. Typically they exhibit the following
general features:
1. Within the useful range of experimental parameters, almost exclusively singly charged ions are detected.
2. To a higher or lesser degree – depending on the experimental
   technique – ion initial energy distributions do not reflect equilibrium distributions and attempts to fit them to a Maxwellian
   distribution leads to temperatures that are too high to be
   realistic.
3. Ions of both polarities are generated, usually at comparable
   abundances. For atomic ions, their yield is governed in first
   order by the ionization potential or electron affinity. For acidic
   compounds, specific molecular ions (e.g. $(M-H)^-$) are found mostly
   in the negative ion spectra; for basic compounds specific ions
   (e.g. $(M+H)^+$, $(M+Alkali)^+$) are more easily identified in the
   positive ion spectra. The somewhat surprising results obtained
   for organic salts will be discussed further down.
4. Relative sensitivity factors for atomic ions out of a given matrix
   vary by only one, in extreme cases by about two, orders of magnitude at least for the high irradiance techniques (1). This holds
   even for ions of only one polarity, i.e. under suitable conditions
   ions such as F^+ or Cl^- are detected in positive ion spectra. This
   contrasts favourably with e.g. SIMS where sensitivity factors
   vary by several orders of magnitude.
5. The spectra of both atomic and molecular ions are qualitatively
   and quantitatively influenced by the surrounding matrix and the
   bonding state of a given species to that matrix. Again this influence becomes less pronounced in the detection of atomic ions
   at higher irradiances.
6. Cluster ions such as $Me^{\pm}$ (5) for metals, $Me_xO_y^{\pm}$ (6, 7) from binary
   metal oxides and $A_xH_g^{\pm}$ (8) for alkali halides, unspecific $C_nH_m^{\pm}$
   clusters as well as specific $(2M+Na)^+$ (9) for organic molecules
   are relatively frequent up to relative cluster masses of several
   hundred. Cluster distributions reflect bond energies and cluster
   stabilities, but they cannot usually be fitted by a distribution

function reflecting a realistic equilibriums temperature (8).
Cluster distributions seem to drastically depend as the sample
preparation technique and the surface state.
For molecular ions some additional features are observed:
7. Polarity of the molecule seems to be supportive to desorption in
    contrast to thermal evaporation, where polar groups must, as a
    rule, be derivatized, to render a molecule volatile.
8. Ions are predominantly of the even electron type, protonation and
    deprotonation rather than electron abstraction or attachment are
    common processes. Radical ions are only rarely generated from
    strongly aromatic compounds (see Fig. 7).
9. Cationization by alkali-metal ions is very frequent. Cationization
    by other metal ions (e.g. Ag, Cu, Mg etc.) has also been observed
    under suitable conditions (10, 11, 12) as has been anionization;
    e.g. by chlorine (11).
10. For the techniques using very short, high irradiance laser pulses,
    a more or less smooth transition to pyrolysis of the sample is
    observed with increasing irradiance. It appears that at least for
    the LAMMA technique this transition does not always occur at iden-
    tical irradiances for positive and negative ions. This will be
    discussed in more detail later in this paper.
        At least for molecular ions, the similarities between all de-
sorption spectra are believed to result from common chemistry that
the molecules or ions undergo. It appears that in all desorption
techniques there must be a step at which such chemical reactions can
occur, most probably as a stabilizing step. Most of this chemistry
seems to be governed by much the same molecular properties that are
also known to govern most of the chemistry in the gas phase or in
solution. It is then obvious that these common features must not
necessarily indicate equal excitation and ion formation processes
for all the desorption techniques or even among the different laser
desorption arrangements. Indeed, the rather different mechanisms,
discussed in the following section, have only rather recently been
recognized, and only after the more subtle differences in the spectra
and other ion properties had been worked out.

Ion formation mechanisms

Four ion formation processes can at least in principle be distinguished. They contribute to varying degrees to the spectra obtained
with the different techniques
1. Thermal evaporation of ions from the solid.
2. Thermal evaporation of neutral molecules from the solid followed
    by ionization in the gas phase.
3. Laser desorption.
4. Ion formation in a laser generated plasma.
        The first two processes are called thermal, because they can
also be induced by classical Joule- or non-laser radiative heating,
usually in conjunction with heat conduction to the sample surface.
Thermal evaporation of cations of quarternary ammonium salts and
anions of sodium tetraphenylborate has been demonstrated by several
groups (13, 14, 15). Such a thermal evaporation of ions, common for
metals and inorganic salts such as alkalihalides, had not originally
been expected to occur for organics as well. It should be most pro-

bable for organic salts with quarternary salts possibly exhibiting a
singularly high yield (13).

Volatility, together with thermal stability are the two terms
that are used to characterize the chance for thermally evaporating
neutral molecules from the condensed into the gas phase. Based on
former experiments employing mostly electron ionization, most organic
molecules, particularly bioorganic ones with relative molecular masses
above about 100 were believed to be involatile and thermally unstable.
Recently Röllgen et.al. (14, 15, 16, 17) as well as Kistemaker et.al.
(18, 19, 20) have published investigations that strongly point to a
thermal evaporation of neutral molecules of sucrose and a number of
other organic substances followed by ionization in the gas phase by
ion-molecule reactions with alkali ions. If no separate source for
the alkali ions is provided, the existence of separate areas of lower
temperature, just high enough for the onset of evaporation of intact
neutral molecules (e.g. ~ 300 - 350°C), and of high enough temperature
(> 750°C) for thermal evaporation of alkali ions are a prerequisite
for this process to occur. In all likelihood, this is the dominating
ion formation process in all the CW-CO$_2$-laser experiments. Besides
ion-molecule reactions in the gas phase, thermal surface ionization
at hot surface areas presumably through decomposition of neutrally
evaporated clusters of salts of carboxylic- and sulfonic acids has
also been suggested (14). COTTER et.al. (22, 23, 24) have in addition
shown that the evaporation of ions and even more of neutrals may per-
sist for times of 10-20 μs, considerably beyond the laser exposure,
if temperature decay is slow. This is because of the amount of energy
deposited and because of limited heat conduction. In the experiments,
pulsed CO$_2$-laser (~ 100 ns) with irradiances around 10^6 W/cm^2 have
been used. Thermal equilibrium should certainly be established on the
molecular scale in this time domain, yet molecular ions of relative
mass beyond several thausand have been observed with peak temperatures,
calculated by the authors, of ca. 2000K.

These results certainly call for future extensive and systematic
investigations of the volatility of organic molecules and ions as a
function of molecular size and structure. There seems to be some
evidence that neutral peptides up to $M_r/Z < 1900$ can be evaporated
by slow radiative heating of samples on a gold substrate (25). In
view of these observations, the frequent statement that polarity
generally supports "desorption" may also need further consideration,
at least in cases where thermal evaporation plays a key role in the
ion formation.

The term desorption is used in contrast to evaporation in cases
in which a transition of a molecular or ion from the condensed into
the gas phase is assumed to take place under non thermal equilibrium
condition. The underlying idea is that at thermal equilibrium, temperatures for an evaporation would lead to a correspondingly high
excitation of internal vibrational modes of excitation leading to
fragmentation of the molecule. As mentioned above, several characteristics of the ion spectra (2., 6.) cannot reasonably be fitted to
an equilibrium temperature model. These properties seem to be the
more pronounced, the higher the laser irradiance (i.e. usually the
shorter the pulse) and are best documented for the LAMMA technique.
Though metastable decay of ions is observed and will be discussed
below, the decay rate for most of the ions is very small and decay

times are long. In particular, peak widths of spectra obtained with
the ion reflector. i.e. with instruments that compensate for the
initial energy distribution, reflect ion generation times of at most
10 ns, i.e. within the laser pulse length. This indicates no appreciable ion decay or gas phase reactions during the ion acceleration
phase; i.e. in the time range of ten to several hundred nanoseconds
at distances beyond ca. 10 μm from the sample surface.

The mechanisms that lead to such laser desorption are now believed to be <u>collective</u>, <u>non equilibrium processes</u> in the condensed
phase (<u>26</u>). In this respect they are closer to processes that must
be assumed to lead to ion generation in SIMS and plasma desorption
rather than to the thermal laser induced ion generation discussed
above, even though the spectra are often indistinguishable for all
different laser techniques. The recently reported observation of
metal ion (Cu, Ag, Mg etc.) attachement for desorption with high
power, short pulse lasers (<u>10, 11, 12</u>) also points to the similarity
with SIMS.

The reason for the dominance of such desorption processes$_R$vs.
thermal evaporation is most easily understood for the LAMMA 500^R
arrangement. Because of the tight focusing of the laser beam, the
total energy delivered is typically only about 100 nJ or less. Moreover, there is no substrate that could absorb energy and then conduct
it to the sample; both sample and thin-film-substrate are usually
"evaporated" simultaneous in the focus during the laser shot. The
total affected volume of about < 0.1 μm^3 expands at a speed of
10^5 cm s^{-1} or more. As a result, probability of gas phase collisions
for a given ion or molecule is very small compared to arragements
with large irradiated areas and/or long irradiation times. Thermal
processes are much more likely for the LAMMA 1000^R arrangement.
FEIGL et.al. (<u>27</u>) were able to show that at ion generation threshold
(. 0.1 μJ) from a bulk metal surface, ion generation occurs essentially during the laser pulse time only. At 10-times threshold energy
(. 1 μJ), the strong ion emission during laser irradiation is followed
by a emission of ions lasting for several microseconds, presumably
of thermal origin due to the heated bulk sample. This is in agreement
with the above discussed results reported by Cotter.

The details of the formation processes of the observed ions are
as yet not well understood. The frequent observation of ions which
have undergone substantial structural rearrangements or even chemical
reactions, unlikely to occur on a nanosecond timescale in the solid
state, suggest that there is an intermediate state of higher mobility,
similiar to suggestions made for SIMS. The upper limits given above
for times and locations within which such reactions would have to
take place would certainly allow such an intermediate state to have
particle densities several orders of magnitude below that of solid
state density. Yet average particle distances and the lack of a
shielding solvent probably prohibit a direct application of liquid-
or gas phase chemistry to this intermediate state, particularly as
it is of transient nature, such that no chemical and probably even no
thermal equilibrium is attained.

At laser irradiances of typically about 10^{10} Wcm^{-2} and above,
dense <u>plasmas</u> are formed from any solid sample, as is well documented
by the large number of laser fusion experiments. In this mode of
operation, energy is deposited into the solid during the initial

phase of the laser pulse only, creating a highly absorbing plasma in front of the surface, shielding it from incoming radiation of the later part of the laser pulse (28). A substantial amount of energy of the laser beam will thereby be deposited into the plasma, increasing its temperature as well as density of ions and electrons. The plasma mode is certainly not suited for organic mass spectrometry. Molecules will be mostly broken down to their atomic constituents, multiply charged atomic ions will become abundant with increasing laser irradiance and ion initial energies will extend into the range of kiloelectronvolts. For the analysis of inorganic specimen such as metals or minerals, this mode has on the other hand been shown to have decided advantages, because the ion yields become nearly uniform for all elements throughout the periodic table (1, 2, 3). It remains puzzling though that even under operational conditions at which the formation of relatively dense plasmas in local thermodynamic equilibriums is expected, abundance of doubly charged ions is usually below the detection limit. This would imply a strong recombination probability prior to acceleration of the ions. Though there seems to be sufficient experimental evidence for the four mechanisms discussed to be active in the various laser desorption experiments, the actual interpretation of experimental results may not be as straightforward. In many cases, more than just one process may contribute to the results observed. For example, it is quite likely that under a suitable choice of experimental parameters in laser desorption with high power lasers, one can simultaneously create a plasma in the center of the non uniformly irradiated area, get laser desorption from the periphery and even thermal emission due to a heated substrate for times longer than the laser pulse.

Discussion

From the above discussion of the characteristics of the spectra and ion formation mechanisms, it is obvious that, though there can be no doubt about the usefulness of laser mass spectrometry for a large variety of analytical tasks, more research is needed for a better understanding. This is particularly true for the transition from thermal evaporation to desorption and the desorption mode itself. In the following, a few first results of such experiments, conducted recently in the author's group, will be reported.

Wavelength dependence of ion formation

Laser wavelength should be one of the key parameters in systematic investigations into ion formation processes. Results should be particularly indicative for bioorganic molecules and wavelengths in the far UV, where strong differences in classical absorption for different groups of such molecules are known to exist. The group at the institute of spectroscopy of the USSR Academy of Sciences has reported strong resonance desorption in the far UV for a number of molecular systems (29, 30, 31). On the other hand, the above mentioned collective, non equilibrium processes do not suggest a direct influence of classical resonance absorption for irradiances above about 10^7 W/cm^2, at which all samples very effectively absorb energy from the radiation field. Nevertheless, the spectra shown in Fig. 1 seem to demonstrate such resonance effects at irradiances close to ion detection threshold. For the wavelength of 266 nm, tyrosine exhibits strong, serine almost no classical absorption. At threshold irradiance, the tyrosine spectrum (top right) shows the parent molecular ion as the base peak with only two other contributions by the decarboxylated molecule and the molecular residue. In particular, no signals of alkali-ions or alkali-attached ions are detected. These ions appear only at irradiance a factor 2 above threshold (lower right). At the same time, the contributions of the fragment-ions increase considerably apparently at the cost of those of the parent molecule. Ion detection for the non absorbing serine occurs only at an irradiance another factor of 3 higher (lower left). Even at threshold strong contributions from alkali ions and alkali-attached molecular ions are always observed, as well as cluster ions. Very similar results have been obtained for other pairs of aromatic and aliphatic aminoacids as well as for small peptides containing only aliphatic aminoacids or those of both types. These experiments will be pursued in more detail in the future.

Formation of metastable ions

As discussed above, metastable decay should, among others, yield information about the degree of internal excitation of the generated ions. A LAMMA 500^R instrument has therefore been supplemented to allow for the simultaneous detection of ions and neutrals (Fig. 2). With the potentials of the various sections of the instrument set as shown, the detector in the ion-reflected path will register all ions at their respective mass numbers that remained stable during the whole acceleration- and flight time ($\sim$ 20 ns $\leq$ t < $\sim$ 50 µs for ions of M_r/Z = 500). Due to the geometry of the reflector, ions created by metastable decay along all of the flight path will usually not reach the detector. Neutrals, formed by decay during the ion flight time in the drift region preceeding the ion reflector ($\sim$ 200 ns < t < $\sim$ 25 µs for parent ions of M_r/Z = 500), will be registered by the second, straight tube detector. It should be noted that these neutrals are recorded at the massnumber of their parent ion, rather than their true mass. With this arrangement, the generation of neutrals through metastable decay and some features of the spectra of positive- vs. those of negative ions out of some organic acids and their salts have been studied.

Figs. 3 and 4 show the positive and negative ion spectra of phthalic acid (up) and their neutrals (down). All major peaks of the spectra of either polarity can be assigned to the parent molecule or to relatively simple fragments. Metastable decay is somewhat more pronounced for the negative ions. It should be noted that the two detection paths have not, so far, been absolutely calibrated relative to each other, so the spectra of ions and neutrals cannot be compared on an absolute scale. The spectra of positive and negative ions of the salt sodium phthalate are shown in Figs. 5 and 6. Though taken at comparable laser irrdiances among each other and to those of the acid, the spectra are quite different from those of the acid. Most of the major peaks in the positive ion spectra can again be assigned to the parent molecule or simple fragments thereof, but metastable decay is considerably more pronounced. The negative ion spectrum is different

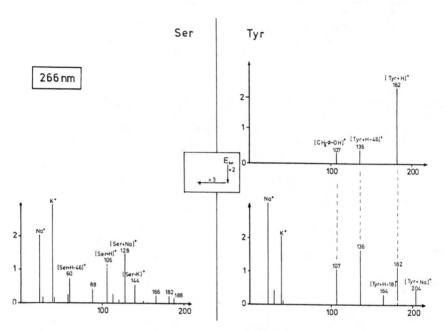

Figure 1.   Spectra of the aminoacids tyrosine (Tyr) and serine
(Ser), obtained at different laser irradiances E_{hv}. Laser wavelength was 266 nm. For explanation see text.

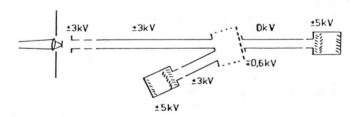

Figure 2.   Schematic diagram of LAMMA 500 instrument, adapted
for simultaneous detection of ions and neutrals.

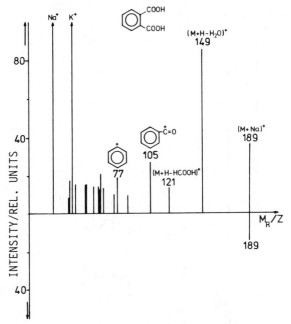

Figure 3. Positive ion spectra (up) of phthalic acid and of neutrals (down) from metastable decay of positive ions.

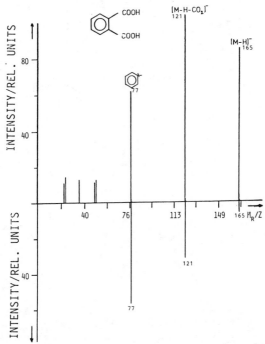

Figure 4. Negative ion (up) spectra of phthalic acid and of neutrals (down) from metastable decay of negative ions.

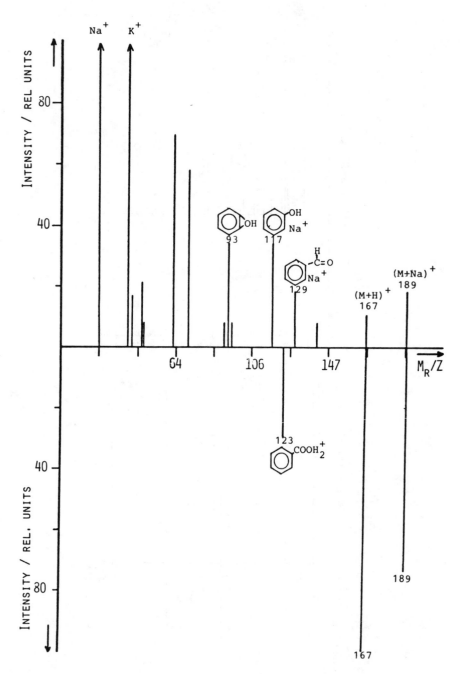

Figure 5.   Positive ion (up) spectra of sodiumphthalate and of neutrals (down) from metastable decay of positive ions.

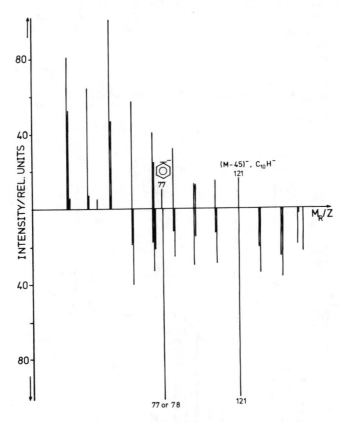

Figure 6.   Negative ion (up) spectra of sodiumphthalate and of neutrals (down) from metastable decay of negative ions. All non assigned peaks represent $C_n H_m$ cluster ions.

altogether. Only the small peak at $M_r/Z = 77$ can unambiguously be assigned to a fragment of the parent molecule; the peak at $M_r/Z = 121$ may contain a contribution from the decarboxylated parent anion, as evidenced by the very strong peak of neutrals. All other ion- and neutral signals represent unspecific C_nH_m-clusters. It is interesting to note further that the centroid of the cluster distribution of the neutrals as compered to the ions is strongly shifted towards larger clusters.

A considerable difference between the spectra of positive and negative ions is also typical for highly aromatic compounds. Fig. 7 shows the spectra of benzopyrene as an example. The positive ion spectrum is amazingly simple. Detection of the radical parent ion is

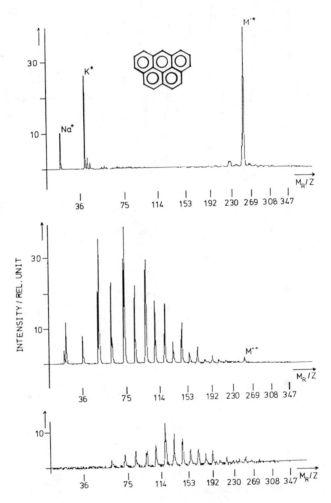

Figure 7.   Positive- (top) and negative ion (middle) spectra and of neutrals (bottom) from metastable decay of negative ions of benzopyrene.

typical for highly aromatic compounds. No neutrals from positive ion decay have been detected. The negative ion spectrum and that of their neutrals on the other hand resemble very much those of the salt shown before. Similar observations as to the differences between positive- and negative ion spectra have also been made by other investigators (32). At this stage of the investigation, we suspect that the phenomenon is observed only for samples, which, upon laser irradiation, generate a substantial number of electrons besides ions and neutrals. In the negative ion mode, these electrons are accelerated along with the ions, resulting in collisions and electron attachement.

The described experiments will have to be continued and other, and more sophisticated experiments need to be designed in order to unravel at least the most important basic processes in laser desorption. Hopefully a better understanding of these processes will then lead to an optimization of the instrumental parameters for a given analysis.

These experiments have in part been supported by the Deutsche Forschungsgemeinschaft and the Bundesministerium für Forschung und Technologie (BMFT).

Literature Cited

1. J.A.J. Jansen and A.W. Witmer: Spectrochimica Acta 37 B (1982)483
2. R.J. Conzemius and H.J. Svec: Anal. Chem. 50 (1978) 1854
3. R.J. Conzemius, F.A. Schmidt and H.J. Svec: Anal. Chem. 53 (1981) 1899
4. F. Hillenkamp in: Ion Formation from Organic Solids, pg. 190-205 A. Benninghoven, ed., Springer Series in Chem. Phys. 25 Springer Pub. Comp., 1983
5. N. Fürstenau and F. Hillenkamp: Int. J. Mass Spectr. Ion Phys. 37 (1981) 135
6. E. Michiels and R. Gijbels: Spectrochim. Acta (1983), 38 B (10), 1347
7. E. Michiels and R. Gijbels: Anal. Chem.
8. B. Schueler, F.R. Krueger and P. Feigl: Int. J. Mass Spectr. Ion Phys. 47 (1983), 3
9. Ch. Schiller, K.-D. Kupka and F. Hillenkamp: Fresenius Z. Anal. Chem. 308 (1981) 304
10. D. Zakett, A.E. Schoen and R.G. Cooks: J. Am. Chem. Soc. 103 (1981) 1295
11. D. Balasanmugam, T.A. Dang, R.J. Day and D.M. Hercules: Anal. Chem. 53 (1981) 2296
12. B. Schueler, P. Feigl, F.R. Krueger and F. Hillenkamp: Org. Mass Spectr. 16 (1981) 502
13. R. Stoll and F.W. Röllgen: Org. Mass Spectr. 16 (1981) 72
14. U. Schade, R. Stoll and F.W. Röllgen: Org. Mass Spectr. 16 (1981) 441
15. R.B. van Bremen, M. Snow and R.J. Cotter: Int. J. Mass Spectr. Ion Phys. 49 (1983) 35
16. R. Stoll and F.W. Röllgen: Z. Naturforsch. 37a (1982) 9
17. R. Stoll and F.W. Röllgne: J. Chem. Soc. Chem. Comm. (1980) 789
18. G.J.Q. van der Peyl, K. Isa, J. Haverkamp and P.G. Kistemaker: Org. Mass Spectr. 16 (1981) 416
19. G.J.Q. van der Peyl, J. Haverkamp and P.G. Kistemaker: Int. J. Mass Spectr. Ion Phys. 42 (1982) 125

20. G.J.Q. van der Peyl, K. Isa, J. Haverkamp and P.G. Kistemaker:
     Nuclear Instr. Methods 198 (1982) 125
21. R.J. Cotter: Anal. Chem. 53 (1981) 719
22. R.J. Cotter and A.L. Yergey: Anal. Chem. 53 (1981) 1306
23. R.J. Cotter and J.-C. Tabet: Int. J. Mass Spectr. Ion Phys. 53
     (1983) 151
24. J.-C. Tabet and R.J. Cotter: Int. J. Mass Spectr. Ion Phys. 54
     (1983) 151
25. E. Constantin and J.F. Muller: 9th Int. Conf. Mass Spectr.,
     paper 14/4, and E. Constantin, personal communication
26. B. Jöst, B. Schueler and F.R. Krueger: Z. Naturforsch. 37a
     (1982) 18
27. P. Feigl, B. Schueler and F. Hillenkamp: Int. J. Mass Spectr.
     Ion Phys. 47 (1983) 15
28. G.G. Devyatykh, S.V. Gaponov, E.E. Kovalev, N.V. Larin, V.I.
     Luchin, G.A. Maksimov, L.I. Pontus and A.I. Suchov: Sov. Techn.
     Phys. Lett. 2 (1976) 356
29. V.S. Antonov, V.S. Letokhov, Y.u.A. Matveyets and A.N. Shibanov:
     Laser in Chemistry 1 (1982) 37
30. V.S. Antonov, V.S. Letokhov and A.N. Shibanov: Appl. Phys. 25
     (1981) 71
31. V.S. Antonov, V.S. Letokhov and A.N. Shibanov: Appl. Phys. B 28
     (1982) 245
32. K. Balasanmugam, S. Viswanadham and D.M. Hercules: Anal. Chem.
     55 (1983) 2424, and D.M. Hercules, personal communication
33. T.P. Hughes: Plasmas and Laser Light, Adam Hilger Pub. Comp.
     Bristol, England, 1975

RECEIVED April 16, 1985

Angle-Resolved Secondary Ion Mass Spectrometry

Nicholas Winograd

Department of Chemistry, The Pennsylvania State University, University Park, PA 16802

The interaction of keV particles with solids has been characterized by the measurement of the angle and energy distribution of sputtered secondary ions and neutrals. The results are compared to classical dynamics calculations of the ion impact event. Examples using secondary ions are given for clean Ni{001}, Cu{001} reacted with O_2, Ni{001} and Ni{7 9 11} reacted with CO, and Ag{111} reacted with benzene. The neutral Rh atoms desorbed from Rh{001} are characterized by multiphoton resonance ionizaton of these atoms after they have left the surface.

The collision of a keV heavy particle with a solid initiates a complex series of events which may ultimately lead to the ejection of a variety of atomic and molecular species. The composition of these species are often characteristic of the original make-up of the target. Whether dealing with a low dose SIMS experiment or a focussed primary beam used for ion microscopy, it is necessary to obtain a detailed understanding of the ion bombardment event to appreciate the mechanisms involved in the ejection process. In a global sense, there are two phenomena which need to be examined. The first is to predict the nuclear motion in the solid which gives rise to the ejection of atoms and the second is to evaluate the inelastic events that give rise to excited species and secondary ions. The first aspect of the problem has been extensively developed utilizing a straightfoward classical dynamics model to follow the flow of energy through the lattice for the first picosecond or so after bombardment.(1) The ionization problem is much more difficult, although progress is now being made using a number of approaches.
      It is our view that in order to compare experimental measurements to the emerging theoretical predictions, it is

0097–6156/85/0291–0083$06.00/0
© 1985 American Chemical Society

necessary to be very careful in specifying the type of sample
to be studied and in defining the measurement conditions as
precisely as possible.

In this paper, we report on a series of experiments aimed at
measuring the yield of secondary ions and neutrals as a function of
their take-off angle and their kinetic energy. This approach, rather
than measuring angle and energy-integrated yields, allows much more
detailed comparisons to theory and makes the testing of proposed
models much more straightforward.

Angular Distributions of Secondary Ions From Clean Single Crystal Surfaces

Using the general experimental and theoretical approach described
above, it is now our goal to see what type of structure-sensitive
information exists in the angular distributions of the secondary
pqns. As Wehner showed many years ago, the distributions of the
neutrals are highly anisotropic and very clearly reflect the surface
symmetry. There have been many attempts to explain these
distributions. One such explanation is that the ejection occurs
along close-packed lattice directions which extend deep within the
crystal.(2) This idea nicely explained the peaks in the angular
distributions but required that there be quite a bit of long range
order in the solid even during the impact event. That requirement
seems a bit hard to swallow in view of the extensive damage that is
created within the crystal. Although controversy existed concerning
these "focusons" for many years, the molecular dynamics calculations
of Harrison clearly showed that the ejection was dominated by near
surface collisions rather than those from beneath the surface.(3)

Ni^+ Ion Angular distribtuions from Ni{001}. The results of
calculations are displayed schematically in Figure 1, where a
Ni{001} crystal face is given as an example. Here, each atom's
ultimate fate is plotted as a point on a plate high above the solid.
Atoms that are ejected perpendicular to the surface ($\theta = 0°$) are
plotted in the center of the plate.

The molecular dynamics calculations yield a clear
picture of the scattering mechanisms that give rise to these
angular anisotropies, particularly for the higher kinetic
energy atoms. Most of the ejected particles arise from within
two or three lattice spacings from the impact point and suffer
only a few scattering events. The spacings between the surface
atoms exert a strong directional effect during ejection. Note
that most particles are ejected along $\phi = 0°$, since there are
no atoms in the surface to block their path. The nearest
neighbor atom along $\phi = 45°$ inhibits ejection in this
direction. It is possible, using the apparatus shown in Figure
2, to compare the measured angular distributions of secondary
ions to the calculated distributions for a clean Ni{001}
single crystal surface.(4) The results of this comparison are
shown in Figure 3.(5) Each panel represents an azimuthal angle
scan at a particular polar angle. The calculated curves have
been corrected for the presence of an image force which tends
to bend the secondary ions toward the surface plane. The
agreement between the two curves is reasonable under all

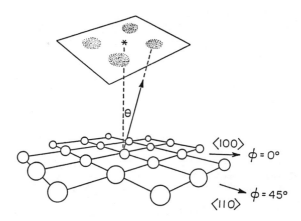

Figure 1. Coordinate system used in determining angular
distributions.

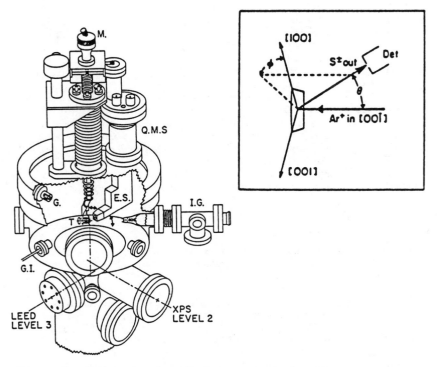

Figure 2. Schematic view of the spectrometer. The components
illustrated include M, crystal manipulator; Q.M.S., quadrupole
mass spectrometer; I.G., primary ion source; E.S., energy
spectrometer; G, Bayard–Alpert gauge; T, crystal target; and
G.I., gas inlet. Auxiliary components are omitted for graphical
clarity. The SIMS experimental geometry and coordinate systems
are defined in the inset. Reproduced with permission from
Ref. 4. Copyright 1981, American Institute of Physics.

conditions. Note that in accord with the schematic
presentation in Figure 1, the secondary ion intensity
maximizes at $\phi = 0°$ and minimizes at $\phi = 45°$ for $\theta \geqslant 45°$. Thus
it appears that in this simple situation,
ion angular distributions behave similarly to the neutrals and are
well-predicted by theory.

Ni_2^+ Ion Angular Distributions from Ni{001}. A real advantage of the
static SIMS method over other surfaces analysis techniques is that
molecular cluster ions may be produced which are characteristic of
the surface chemistry. It would be most interesting, then, to be
able to obtain the angular distribution of these species to see if
they too contain information about surface structure. As shown in
Figure 4, the experimental angular distributions for Ni_2^+ ions
actually exhibit a sharper azimuthal anisotropy than the Ni^+
ions.(5) This result is also observed in the classical dynamics
calculations.(6) Of particular interest in this case is the fact
that the mechanisms that give rise to this increased anistropy can
be ascertained from the theory. As it turns out, most of the dimers
that are formed at $\phi = o°$ and $\theta = 45°$ originate from a similar set
of collision sequences as illustrated in Figure 5. The Ar^+ ion
strikes target surface atom 4, initiating motion in the solid that
eventually ejects atoms 1 and 3 into the vacuum. Both of these atoms
are channeled through the fourfold holes in the $\phi = 0°$ direction,
are moving parallel to each other, and are fairly close together.
Note that the two atoms that form the dimer do not originate from
nearest-neighbor sites on the surface.
        There is an important ramification of the concept that the
dimers that give rise to the maxima in the angular distribution are
formed primarily from constituent atoms whose original relative
location on the surface is known. If this result were extrapolated to
alloy surfaces such as $Ni_3Fe(111)$, the relative placement of the alloy
components on the surface would be determined. For example, for the
$Ni_3Fe(111)$ spectra there should be no nearest neighbor Fe atoms on a
perfect (111) alloy surface, yet an Fe_2^+ peak is observed.(7)

Angular Distributions of Secondary Ions From Adsorbate Covered
Surfaces

The channeling phenomena observed from clean surfaces should
also be found in more complex systems such as metals covered
with a chemisorbed layer. For these cases, there are various
ways in which one might envision the angle to be important.
Examples of azimuthal anisotropies have already been seen for
the case of clean metals where surface channeling and blocking
give rise to the observed effect. This situation should also
apply to adsorbate covered surfaces. Other possibilities
include the study of anisotropies in the polar angle
distributions as well as in the yield of particles due to
changes in the angle of incidence of the primary ion.
        Considerable progress in quantitatively describing the
ejection of chemisorbed atoms and molecules from metals has
been made using molecular dynamics calculations. The main
difficulty in describing any situation like this is to develop
appropriate interaction potentials which describe the

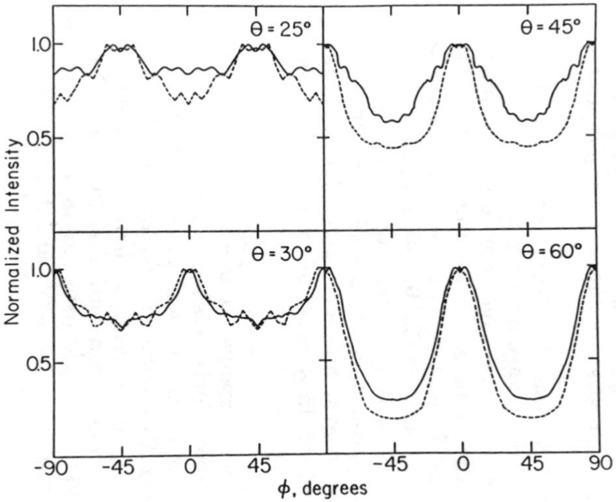

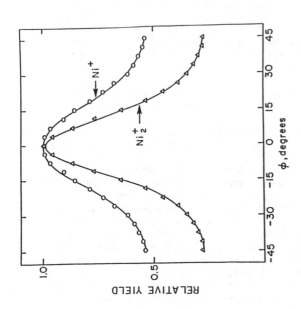

Figure 4. Experimental angular distributions of Ni+ and Ni2+ ejected from Ni(001) at a polar angle of 45±5°. The center-of-mass kinetic energy of the particles is between 10 and 50 eV. Both curves are fourfold averages of the kinetic energy and is at normal incidence. The solid is at room temperature. The peak counts are ≈900 and ≈500 counts/sec for the Ni+ and Ni2+ distributions, respectively. The <100> azimuthal directions correspond to φ=0°. Reproduced with permission from Ref. 6. Copyright 1980, American Physical Institute.

Figure 3. Dependence of Ni+ ion yield on azimuthal angle at various polar angle for clean Ni(001) bombarded by 1500 eV Ar+ ions at normal incidence. The solid curves represent experimental data while the dashed curves are predicted values obtained by correcting the calculated yields for 1000 eV Ar+ ion bombardment for the presence of the image force. Only those particles with a kinetic energy of 4±4 eV were detected. Reproduced with permission from Ref. 5. Copyright 1982, American Institute of Physics.

scattering events. Since little is known about these potentials, early calculations have utilized pair-wise additive potentials for adsorbates which have the same form as for the substrate, but with different mass. The exact form of the potential is not as critical as the atomic placement of the adsorbate atom. Thus, in the calculation, the geometry and coverage of the adsorbate may be varied over a wide range to test how these quantities influence ejection mechanisms and ultimately the angular distributions. In this section, examples of how several different experimental configurations can be utilized will be reviewed.

Atomic Adsorbate. The first application of angle-resolved SIMS to the determination of the surface structure of chemisorbed layers is for oxygen adsorbed on the {001} face of Cu.(6) In this situation, the oxygen overlayer forms a c(2x2) structure as determined by LEED. Classical dynamics calculations indicate that the oxygen should be ejected in the $\phi = 0°$ direction if it is originally bonded above the copper atom, because it is directly in the path of the ejected substrate species. However, if the oxygen is in a hole site, bonded to four substrate atoms, its predicted angle of ejection is $\phi = 45°$. Experimental studies have confirmed that the oxygen resides in a fourfold bridge site because it is ejected in the $\phi = 45°$ direction.(6)

There are a number of complications associated with this simple interpretation. First, the magnitude of the azimuthal anisotropies are dependent upon the kinetic energy of the desorbing ion. For the very low energy particles, there has been sufficient damage to the crystal structure near the impact point of the primary ion that the channeling mechanisms are no longer operative. On the other hand, at higher kinetic energies, say greater than 10 eV, the desorbing ion leaves the surface early in the collision cascade while there is still considerable order in the crystal. The channeling mechanisms are much stronger and the angular anisotropies are larger.

A second complication involves the determination of the height of the adsorbate atom above the surface plane. Calculations have been performed where this bond distance has been varied over several angstroms in order to find the best fit with experiment.(8,9) These studies have also shown that there is a sensitivity of the polar angle distribution to the effective size of the adsorbed atom. Thus, it is important to know more about the scattering potential parameters if this distance is to be determined accurately. It appears, however, that the type of adsorption site may be determined in a reasonably straightforward manner.

Adsorption of CO on Ni{001}. The response of a surface to ion bombardment covered with a molecularly adsorbed species is mechanistically distinct from the atomic absorbate case. For CO on Ni{001}, for example, the strong C–O bond of 11.1 eV and the weak Ni–CO bond of 1.3 eV allows the CO molecule to leave the surface without fragmentation. In the experimental studies, the main peaks in the SIMS spectra for a Ni{001} surface exposed to a saturation coverage of CO are Ni^+, Ni_2^+, Ni_3^+, $NiCO^+$, Ni_2CO^+, and Ni_3CO^+. All ions show a smooth increase in intensity with CO adsorption and reach saturation after 2-L CO exposure (0.5 monolayer coverage). The yields of C^+, O^+, NiC^+ and NiO^+, are all less than 0.01 of the Ni^+ intensity. The classical dynamics treatment for CO on Ni{001} yields results which are in qualitative agreement with these findings. Approximately 80% of the CO molecules that eject are found to eject intact, without rearrangement. The formation of NiCO and Ni_2CO clusters have been observed to form over the surface via reactions of Ni atoms and CO molecules. No evidence has been found for NiC and NiO clusters in the calculations. The ion bombardment approach, then, is a very sensitive probe for distinguishing between molecular and dissociative adsorption processes.

A number of workers have attempted to identify structural relationships found using other techniques such as LEED and vibrational spectroscopy to cluster yields in SIMS. The correlation of Ni_2CO^+ to bridge-bonded CO and $NiCO^+$ to linear bonded CO is an example of this approach. As it happens, the calculations clearly show that the mechanism of cluster formation is not consistent with this picture since the clusters form over the surface via atomic collisions. Furthermore, combined LEED/SIMS results indicate that the cluster ion yields are not directly related to the adsorbate/substrate geometry.(10) The c(2x2) structure of CO on Ni{001} with all the molecules in the atop site gave the same $Ni_2CO^+/NiCO^+$ ratio as the compressed hexagonal LEED structure which must have both A-top and bridge-bonded CO molecules.

On the other hand, it is clear that angular distributions for atomic adsorbates are very sensitive to the surface structure so it is not unreasonable to anticipate similar effects for the Ni/CO system. Extensive calculations using the molecular dynamics procedure(5) have been completed for the atop and twofold bridge bonding configurations but statistical considerations have restricted the analysis to only the Ni atoms. As shown in Figure 6 when the CO is in the A-top geometry, the calculated Ni distributions peak along azimuthal directions which are similar to the clean surface. For the twofold bridge case, however, the CO overlayer tends to randomly scatter the ejecting Ni atoms producing a much different pattern.(11) The predictions for the atop bonding geometry, when corrected for the presence of the image force, are in quite good agreement with experiment, and are consistent with the wide range of other experimental data available for the system.

Adsorption of CO on Ni{7 9 11}. Since the azimuthal angle distributions are sensitive to subtle differences between surface structures, it is of interest to examine the role of larger surface irregularities such as surface steps on the measured quantities. For example, suppose the orientation of the primary ion beam in the SIMS experiment is fixed at different azimuthal angles with respect to the step edge. If the ejection process is structure-sensitive, then changes in yield and cluster formation probabilities should be observed as the ion bombards "up" or

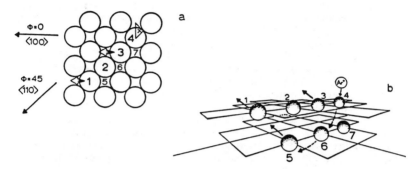

Figure 5. Mechanism of formation of the Ni_2 dimer which
preferentially ejects in the <100> directions contributing the
majority of intensity to the peak in the angular distribution.
(a) Ni(001) showing the surface arrangements of atoms. The
numbers are labels while the X denotes the Ar^+ ion impact point
for the mechanism shown in Figure 8b. Atoms 1 and 3 eject as
indicated by the arrows forming a dimer, which is preferentially
moving in a <100> direction. (b) Three-dimensional representation
of a Ni_2 dimer formation process. The thin grid lines are drawn
between the nearest-neighbor Ni atoms in a given layer. For
graphical clarity, only the atoms directly involved in the mechanism
are shown. Reproduced with permission from Ref. 6. Copyright 1980,
American Physical Institute.

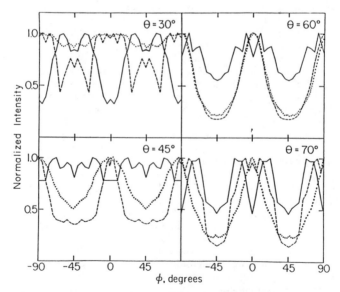

Figure 6. Predicted Azimuthal dependence of the Ni^+ ion yield for
Ni(001)c(2x2)-CO for CO adsorbed in atop (---) and twofold sites
(———). Only those particles with kinetic energies of 3 ± 3 eV
were counted. Experimental points (····). Reproduced with
permission from Ref. 11. Copyright 1984, Springer-Verlang.

"down" the steps. In addition, the desorption of chemisorbed
molecules should be influenced by their proximity to the step
edge.

Carbon monoxide chemisorption on Ni{7 9 11} represents an
interesting case with which to check these concepts since
comparable studies have been performed on Ni{001} and Ni{111} and
since a number of other experimental methods have been applied to
this system. Electron energy loss spectroscopic (EELS) studies
performed at 150 K suggest that the initial adsorption occurs in
threefold and twofold bridge sites along the step edge. Beyond
this point, the CO molecules begin to occupy terrace sites.(12)
Thus, the low temperature adsorption of CO on Ni{7 9 11} presents
a realm of interesting structural phases which should be
sensitive to the azimuthal angle of incidence of the primary ion
beam.

The experimental results for the NiCO⁺ ion yield as a
function of angle is illustrated for this system in Figure 7 and
the angles are defined in Figure 8. Note that the cluster ion
yields are higher at ϕ = 180° than at ϕ = 0°, with the most
significant variations occurring at intermediate angles. At 0.2 L
exposure, the NiCO⁺ ion signal shows a broad peak which appears
at ϕ = 115°. This peak shifts slightly to 105° and sharpens
somewhat at an exposure of 0.4 L. By 0.6 L exposure the peak has
become very intense and is only 10° wide, centered at ϕ = 100°.
As the CO coverage increases this peak becomes very broad. At the
saturation exposure of 2.8 L the NiCO⁺ ion intensity displays a
broad maximum between ϕ = 40° and ϕ = 160°. An exposure of 0.6 L
corresponds almost exactly to the exposure at which the EELS
results indicate that all the CO molecules were bound to
adsorption sites near the step edge and that all the edge sites
were occupied.(10) Apparently, the specific bonding site of the
CO next to the step edge is responsible for the sharp peak in the
NiCO⁺ ion signal at ϕ = 110°. At saturation, the peak loses this
definition completely, presumably since the CO molecules occupy
several sites. At CO exposures performed at room temperature the
azimuthal plots do not exhibit such sharp features, as
illustrated in Figure 7. Calculations performed for the twofold
bridge step-edge adsorption geometry corresponding to the 0.6 L
exposure point successfully reproduce the sharp feature at ϕ =
120°, although it has not yet been possible to identify the
specific collision mechanisms that cause it to occur.(13)

Angle-Resolved SIMS Studies of Organic Monolayers. We have seen
how the angular distributions reflect the bonding geometry of
adsorbates through analysis of the azimuthal anisotropies and by
varying the angle of incidence of the primary ion. The next
possibility is to see if there are channeling mechanisms which
act perpendicularly to the surface and which manifest themselves
in the polar angle distributions. The model systems which
illustrate this effect are benzene and pyridine adsorbed on
Ag(111) at 153 K. These model systems are of interest for a
number of reasons. (i) The molecules are similar in size and
shape and should behave in a closely related fashion under the
influence of ion bombardment. (ii) Classical dynamics
calculations have been performed on these molecules adsorbed on

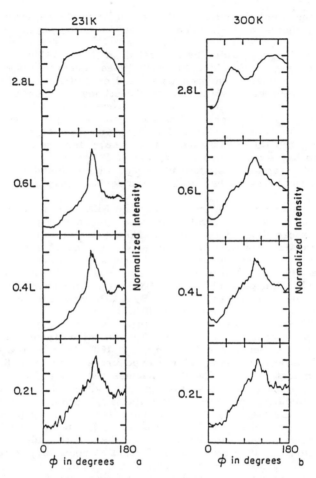

Figure 7. Normalized NiCO$^+$ intensity versus azimuthal angle ϕ
as a function of CO exposure. Reproduced with permission from
Ref. 13. Copyright 1984, American Institute of Physics.

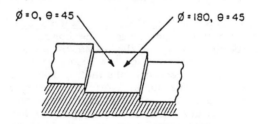

Figure 8. Definition of the polar (θ) and azimuthal (ϕ) angles
of incidence of the primary ion beam relative to the Ni$\{7\ 9\ 11\}$
surface. Reproduced with permission from Ref. 13. Copyright 1984,
American Institute of Physics.

Ni(001) where dramatic differences in the molecule yield are predicted to occur with molecular orientation.(14) (iii) Electron energy loss spectroscopy indicates that pyridine on Ag(111) initially adsorbs in π-bonded configuration but undergoes a compressional phase transition to a σ-bonded configuration as the coverage is increased.(15) Benzene, on the other hand, is believed to remain in the π-bonded configuration at all coverages.(16) A more detailed discussion of these effects is presented in reference 1.

Angular Distributions of Neutral Atoms Desorbed From Single Crystal

Most experimental studies aimed toward determining the angular distributions of secondary particles have focussed on the measurement of the secondary ions. The reason for this emphasis is that there have been no techniques available for detecting the neutral species with monolayer sensitivity. It would be extremely valuable to be able to perform these experiments to be able to obtain data that was directly comparable to the classical dynamics calculations and to get some insight into how the secondary ion fraction is affected by the take-off angle.

In this section, we describe a new apparatus and some preliminary results, aimed at providing detailed trajectory information on the ejected neutrals. It is based on a time-of-flight measurement for the neutral energies, multiphoton resonance ionization (MPRI) for the particle selectivity,(17) and two-dimensional position sensitive detection for the angular information. The detector is operated in an ultra-high vacuum environment, on well-characterized surfaces, and with low primary ion dosages onto the sample. A schematic representation of the experiment is illustrated in Figure 9. The desorption is initiated by a 0.2 μs, 5 KeV Ar$^+$ ion pulse incident on the sample at 45° focussed to 0.2 cm^2, and ionization is accomplished by absorption of photons from a 5 ns laser pulse obtained from the output of a Nd:YAG pumped dye laser. Under the present operating conditions we can detect neutrals whose kinetic energies vary from 0.2-50 eV into a total enclosed angle of over 100°. A complete analysis may be performed using a total dose of less than 10^{12} incident Ar$^+$ ions/cm^2. A detailed description of the apparatus will be given elsewhere.(18)

Using this detector, we have initiated a series of experiments aimed at determining the energy and angular distributions of Rh atoms ejected from clean and adsorbate covered polycrystalline and single crystal surfaces. Rhodium atoms may be efficiently and selectively ionized using 312.4 nm laser light, obtained by frequency doubling the output of the dye laser. From the polycrystalline material, we find the velocity distribution of Rh atoms follows closely the form predicted by Thompson(19) with a peak intensity occurring at ~5 eV and a high energy tail decreasing in intensity as E^{-2}. Polar angle distributions exhibit nearly a cos^2 shape. From a Rh{001} crystal, the velocity distribution generally peaks at a higher value than that found from the polycrystalline surface, and depends strongly on the value of the polar collection angle. For example, the energy of the emitted atoms tend to be distributed about higher kinetic energies when the polar angle is chosen to coincide with a peak in the atom intensities, a result in qualitative agreement with classical dynamics calculations.

In addition to energy distribution measurements into a given angle, we are able to extract angular distribution measurements of particles with a given energy. Polar distribution measurements at a given azimuth from Rh{001} show three peaks of preferred ejection angles. The position of these peaks are predicted well by the classical dynamics calculations as shown in Figure 10. Of particular interest is the peak observed normal to the surface. This normal ejection peak is more prominent at 30 eV than at 10 eV which corresponds to an energy distribution with a larger high-energy tail. Variations in the relative intensity of this center peak relative to the side peaks are observed when an absorbate such as sulfur is placed on the crystal surface. A preliminary example of this effect is shown in Figure 11. It is hoped that these variations, when coupled to computer simulations of the ion impact event, will lead to a new approach for characterizing such adsorbates.

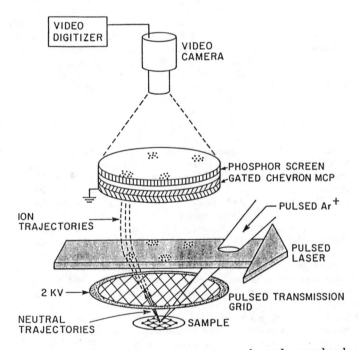

Figure 9.   Detector for performing energy and angle-resolved measurement of neutrals desorbed from surfaces.

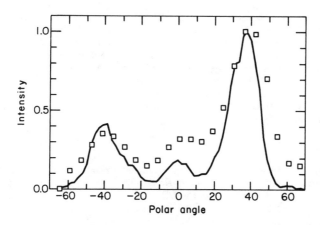

Figure 10. Measured angular distributions from clean Rh{001}.
Rh atoms with kinetic energies between 2 and 34 eV
are collected. The points represent experimental
data while the line represents calculated results.
The crystal is aligned along ϕ = 0°.

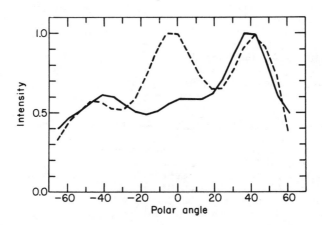

Figure 11. Measured angular distributions for clean Rh{001}
(———) and Rh{001} covered with approximately 10%
of a monolayer of sulfur (----). Rh between 2 and 8
eV are collected. Surface cleanliness as coverages
were estimated using Auger spectroscopy.

Conclusions

It is hoped that the experimental examples discussed in this
paper, together with the numerous comparisons to classical
dynamics calculations, yield convincing evidence that the ion
bombardment phenomenon is becoming well-understood. Of particular
interest is the fact that the trajectories of the secondary ions
appear to follow reasonably closely those of the calculated
neutrals. In addition, angle and energy-resolved measurements may
provide a new approach for elucidation of the structure of single
crystal surfaces.

Acknowledgments

The author is grateful for the financial support of the National
Science Foundation, the Office of Naval Research and the Air
Force Office of Scientific Research.

Literature Cited

1. Garrison, B. J., see her earlier article in this volume.
2. Silsbee, R. H. *J. Appl. Phys.* 1957, 28, 1246.
3. Harrison, D. E.,Jr., Moore, W. L. and Holcombe, H. T. *Radia. Eff.*,
   1973, 17, 167.
4. Gibbs, R. A. and Winograd, N. *Rev. Sci. Instrum.*, 1981, 52, 1148.
5. Gibbs. R. A., Holland, S. P., Foley, K. E., Garrison, B. J., and
   Winograd, N. *J. Chem. Phys.* 1982, 76, 684.
6. Holland, S. P., Garrison, B. J., and Winograd, N. *Phys. Rev. Lett.*,
   1980, 44, 756.
7. Bleiler, R. J., Diebold, A. C., and Winograd, N. *J. Vac. Sci.
   Tech.*, 1983, A1, 1230.
8. Holland, S. P., Garrison, B. J., and Winograd, N. *Phys.
   Rev. Lett.*, 1979, 43, 220.
9. Kapur, S. and Garrison, B. J. *J. Chem. Phys.*, 1981, 75, 445.
10. Hopster H. and Brundle, C. R. *J. Vac. Sci. Technol.*, 1979, 16,
    548.
11. Winograd, N., *Chemical Physics,* 35, Springer Series in
    1984, p.403.
12. Erley, W., Ibach, H., Lehwald, S. and Wagner, H.
    Surf. Sci., 1979 83, 585.
13. Foley, K. E., Winograd, N., Garrison, B. J. and Harrison,
    D. E., Jr., *J. Chem. Phys.,* 1984, 80, 5254.
14. Garrison, B. J., *J. Am. Chem. Soc.,* 1982, 104, 6211.
15. Demuth, J. E., Christmann, K., and Sando, P. N. *Chem. Phys.
    Lett.,* 1980, 76, 201.
16. Friend, C., and Muetterties, E. L., *J. Am. Chem. Soc.,* 1981, 103,
    773.
17. Kimock, F. M., Baxter, J. P., Pappas, D. L., Kobrin, P. H., and
    Winograd, N. *Anal. Chem.,* 1984, 56, 2982.
18. Kobrin, P. H., Baxter, J. P., and Winograd, N. *Rev. Sci. Instr.,*
    in preparation.
19. Thompson, M. W. *Phil. Mag.* 1968, 18, 377.

RECEIVED April 16, 1985

Secondary Ion Mass Spectrometer Design Considerations for Organic and Inorganic Analysis

C. W. Magee

RCA Laboratories, Princeton, NJ 08540

This paper describes, in detail, the various instrument design aspects which must be considered when building a secondary ion mass spectrometer for organic and inorganic analyses. The nature of the information desired from these two kinds of analysis is very different, so it is not surprising that the secondary ion mass spectrometers specificially designed for organic and inorganic analyses are also quite different. Major component areas such as primary ion beam, secondary ion optics, mass spectrometer, detection and vacuum systems are discussed.

The technique of secondary ion mass spectrometry (SIMS) is one of extremely large scope. By using energetic particle bombardment of a surface and detecting with a mass spectrometer the charged particles which are emitted, one can
(1)   detect elements or molecules;
(2)   detect ions ranging in mass from 1 to >5000 Daltons
(3)   detect concentrations of constituents ranging from 100% to a part-per-billion
(4)   generate the detected signal from the first monolayer of the sample surface or sputter to depths greater than 10 um below the original surface
Not surprisingly, however, all of these capabilities are generally not achieved within a single SIMS instrument. It is necessary to examine critically the requirements at hand and choose those design features which lead to an instrument best suited to the defined need.

The features from which one has to choose in building a SIMS instrument are as numerous and varied as the problems which the technique can address. Shown conceptually in Figure 1 are the various choices for primary bombarding beam formation and emitted particle detection, all encompassed within a vacuum system.

The ion source can be of a number of different types capable of making ions of a large variety of atoms. One may also bombard the surface with a beam of energetic neutral atoms or a photon beam

0097-6156/85/0291-0097$06.00/0

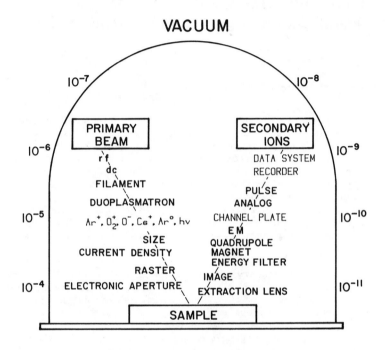

Aldehyde/Ketone Derivatization

Figure 1. Schematic representation of a secondary ion mass
spectrometer.

generated by a laser. The beam size and current density are important, and depending on the application, one may wish to raster the primary beam and use electronic signal gating techniques to spatially select the area from which to accept the signal. The sample may be a bulk solid, thin film, monolayer, or a liquid.

The system for mass analysis and detection of the emitted ions also involves many choices of components. In order to transfer ions most efficiently from the sample to the mass spectrometer, one may choose to use an ion extraction lens. Depending on requirements, this lens may or may not be of the image-forming immersion type. Most instruments will require some kind of energy filter to select the kinetic energy of the ions entering the mass spectrometer. The mass spectrometer itself may be a quadrupole, magnetic sector, or time-of-flight instrument with either a conventional electron multiplier or channel plate detector. The detector system can be operated in either the analog mode or the pulse counting mode with data handling ranging in complexity from a strip chart recorder to a computer data acquisition system.

Of course, everything except the electronics and data system associated with the SIMS experiment must be in a vacuum system, with the experimental results desired dictating the degree of vacuum required. This can range from ultra-high vacuum to pressures achievable with a simple diffusion pump.

System Requirements for Inorganic and Organic SIMS

When designing a SIMS instrument to give the best possible results for a given type of analysis one must know which components are necessary for the desired performance, and indeed, what kind of performance is required. Table I lists the three basic areas of SIMS analysis to be discussed here. Under each type of analysis are listed the most appropriate components and design features.

Inorganic SIMS--Depth Profiling

Inorganic secondary ion mass spectrometry deals with the analysis of the elemental composition of a sample. Due to the fact that the sample must be sputtered away to generate a SIMS signal, the instantaneous sample surface from which the secondary ions are emitted continuously recedes into the bulk. If ions of interest are monitored as a function of analysis time, depth profiles of those elements in the solid are obtained. This is the analysis method most often used in inorganic SIMS. The requirements of this method of analysis dictate the design of the inorganic SIMS instrument. The quality of the primary ion beam of a depth profiling inorganic SIMS instrument is very important. It must be of high current density. In order to reduce crater edge effects,(1) the beam must be rastered 5 to 10 beam diameters on a side to produce a flat-bottomed crater from the center of which the data are taken. Ions generated from the sloping crater wall are rejected by an electronic aperture (2) or by the ion optics of the secondary ion extraction lens (discussed later). In addition, it is important that no energetic neutrals be allowed to hit the sample because they will create ions from the sloping crater walls (3,4) and these

Table I. Instrumental Requirements for Inorganic and Organic SIMS

	Inorganic SIMS Depth Profile	Organic SIMS SOLID Surface	Organic SIMS LIQUID Surface
PRIMARY BEAM	>High Current Density >10 - 100um Diameter >Rastered >No Neutrals >Mass Analyzed	>Low Current Density >100um Diameter >Rastered >No Neutrals	>High Current Density >100um Diameter >Non-Rastered >Ions or Neutrals
SECONDARY ION OPTICS	>Immersion Lens >Electrostatic Analyzer	>Immersion Lens >Electrostatic Analyzer >Dynamic Emittance Matching	>Immersion Lens >Electrostatic Analyzer
MASS SPECTROMETER	>Large Quad or Sector >High Abundance Sensitivity >Mass Range 1 - 240 >High Mass Resolution	>High Mass Quad or Sector >Mass Range 10 - 1000 >Low Mass Resolution	>High Mass Quad or Sector >C.I.D. >Mass Range 60 - 5000 >High Mass Resolution
DETECTION SYSTEM	>Pulse Counting >Data System	>Pulse Counting >Chart Recorder	>Analog >Data System
VACUUM	>UHV	>UHV	>HV
OPERATIONAL MODE	>Scanning >Selected Ion Monitoring	>Scanning	>Scanning

ions will not be rejected by electronic gating. Mass analysis of
the primary beam is generally desirable to insure that no impurities are present in the beam. Such species will become implanted
into the sample and preclude their analysis at low concentrations
in a sample (2).

Once secondary ions are generated from the analysis area of
interest, it is essential to transmit them from the sample surface
to the entrance slit or aperture of the mass spectrometer. This is
most often done with an immersion lens, so named because the sample
is "immersed" in the electric field of the lens (2,5). An electrostatic analyzer (ESA) is also needed to select an energy bandpass
of secondary ions for the mass spectrometer. This is needed in
inorganic SIMS because atomic ions emitted from the surface have an
energy spread of several tens of electron volts. In addition, an
ESA serves to block the line-of-sight from the sample surface to
the mass spectrometer making it impossible for backscattered
primary ions and emitted photons to generate a background signal in
the electron multiplier ion detector.

The mass spectrometer used in inorganic SIMS can either be a
magnetic sector instrument or a quadrupole instrument. It need
only have a mass range of 1 to 240 in order to measure all atomic
ions in the periodic table. This small mass range is beneficial
because it allows one to use a quadrupole rod structure with large
diameter rods thus increasing ion transmission. Alternatively,
with a sector-type instrument one can use a high accelerating
potential which also improves ion transmission. For many materials
problems it is also desirable to have a mass spectrometer capable
of high mass resolving power for separation of closely spaced peaks
in the mass spectrum which occur at the approximately same nominal
mass (6).

The detector system used in most inorganic SIMS instruments
consists of a discrete dynode electron multiplier feeding ultrasen-
sitive, high speed pulse counting electronics. A charge sensitive
amplifier capable of detecting 10^4-10^5 electrons at 100 MHz (7)
should be used along with an electron multiplier capable of
delivering count rates >10 Mc/s with no gain degradation and with
low background (<0.1 c/sec). The 6-7 order-of-magnitude range in
signal intensity observed in inorganic SIMS requires such elaborate
ion detection schemes. One must also consider the vacuum required
in the sample chamber during analysis.

Ultra-high vacuum is desirable in the vicinity of the sample
in order to keep the sample "clean" during analysis. Molecules
from the residual gas strike the sample at a rate equivalent to one
monolayer/sec at 10^{-6} torr. Elements striking the sample from the
residual gas such a H, C, O and N are analyzed as though they were
originally present in the sample thus yielding an anomalously high
concentration for these species.

Organic SIMS - Solid Surfaces

Organic SIMS, as the name implies, deals with the sputtering and
ionization of organic molecules from surfaces. The ability to
detect sputtered ions of organic molecules was first reported by
Benninghoven (8) who studied the emission of intact amino acid ions

from a solid surface. An instrument suited for such analyses
evolves quite naturally from an inorganic SIMS instrument. As
Table I shows, many of the design requirements are the same yet
there are several important differences which make this type of
machine very special.

Of primary importance when sputtering organic molecules is
that those molecules have many interatomic bonds which can be
broken easily by the impact of the primary bombarding particle. In
order to prevent the sputtering of severely damaged molecular
fragment ions of little analytical significance, it is necessary to
limit the experiment such that ions are emitted only from regions
of the sample which have not been damaged by previous primary
particle bombardment. Practically speaking, this generally means
that a complete spectrum must be taken before 10% of the uppermost
monolayer of sample has suffered radiation damage. Otherwise, the
probability of ion emission from a previously damaged region will
be appreciable (9). One can understand readily that this type of
analysis is extremely sample limited (10% of one monolayer). In
order to obtain the largest signal possible, or to obtain a useable
signal for as long a time as possible, one must design an instrument which sputters and extracts ions from as large an area of the
sample as possible. It is this requirement that sets this kind of
SIMS instrument apart from all others.

For optimum performance, an instrument designed to detect
organic ions from solid surfaces should use a primary ion beam of
low current density and relatively small size. The beam should not
contain any energetic neutral particles, and it should be rastered.
The reasons behind these requirements are intimately related to the
need for large-area secondary ion generation and extraction and
will be covered in the discussion. For analysis of nonconducting
surfaces, a beam of energetic neutral atoms may have to be employed
to reduce sample charging. But due to the extremely low fluences
used in organic surface layer analysis, ion beams will usually not
result in severe sample charging.

The secondary ion optics required for large-area ion extraction are very specialized and will also be covered in the discussion. They involve an immersion lens and dynamic matching of the
ion emission point with the entrance aperture/slit of the mass
spectrometer. Again, an ESA is needed, more however, to filter out
high-energy neutral particles and photons than to energy-filter the
secondary ions because organic molecular ions are emitted with a
characteristically low spread of kinetic energies (9).

The mass spectrometer used in organic SIMS of solid surfaces
may be a quadrupole, magnetic sector, or time-of-flight type, but
should preferably have a mass range to approximately 1000 D. Low
mass resolving power is used because the signal levels are generally too low for use with high mass resolving power spectrometers
which have lower ion transmission factors.

Like their inorganic counterparts, solid-surface organic SIMS
instruments use single-ion detection schemes except that the high
count rate requirements are not so severe. There is, however, the
possibility of using a magnetic sector spectrometer with Mattauch-
Hertzog geometry and photoplate or microchannel plate detection.
This is advantageous because all masses are collected simultaneously, thus making maximum use of the secondary ions generated.

The vacuum requirements for this type of SIMS instrument are extremely stringent because the uppermost monolayer of the sample must not become contaminated by the residual gas in the instrument, otherwise the spectrum observed will bear little resemblance to the actual sample surface.

Organic SIMS - Liquid Surfaces

Several years ago, a novel sample preparation technique was developed (10) for sputtering intact organic molecules from surfaces. The key feature of the technique is the use of a viscous, non-volatile liquid surface from which to sputter and generate organic secondary ions.

The actual momentum transfer process involved in sputtering (11) is the same for liquids and solids due to short time duration (picoseconds) of a collision cascade during which the atoms of a liquid are essentially motionless (9). However, during the time between successive impacts of primary particles on a given $100\text{\AA}^2$ area on the sample surface (approximate area damaged by primary particle impact), the molecules of a liquid can move considerable distances. Typical molecular velocities are 30 m/sec. (9) This means that as fast as the uppermost surface monolayer is damaged by primary particle bombardment, it is mixed into the liquid bulk by diffusion and replaced by new, undamaged material. This liquid-phase mixing allows large fluxes of primary bombarding particles to be used while still maintaining an undamaged surface monolayer from which to emit secondary ions. The large bombarding flux will, in turn, produce a large flux of secondary ions, large enough to be used by a more conventional organic mass spectrometer.

As just mentioned, a liquid surface organic SIMS instrument should use a high flux primary bombarding beam. It may be an ion beam or a beam of energetic neutral atoms. In either case, it should have a small diameter for reasons which will be made clear in the discussion. Care must be taken, however, to keep the current density of the bombarding beam within reasonable limits (several mA/cm^2) in order to keep within this damage/mixing regime. Ion sources of exceeding high brightness, such as field emission guns, can produce ion beams capable of being focussed to current densities approaching 1 A/cm^2. Under these conditions, the time between successive primary ion impacts within the $100\text{\AA}^2$ area of damage would be so short that the mixing of the damage into the bulk could not occur from one primary ion impact to the next.

The secondary ion extraction optics should include an immersion lens above the sample surface to maximize secondary ion transmission from the sample surface to the mass spectrometer. An electrostatic analyzer should again be used to filter out high-energy neutral particles and photons from the secondary ion beam. Energy analysis of the secondary ions is generally not necessary, but can be used to vary the amount of fragmentation observed in the mass spectrum (9).

The rest of the mass spectrometer can be a conventional organic instrument. High mass quadrupole mass filters or magnetic sector instruments may be used. A mass range to 5000 D is advis-

able to make use of the measurable ion currents of very large molecular ions of biomedical importance which can be generated using
the liquid-matrix techniques. One may also employ high mass resolution and collosion induced-dissociation techniques due to the
large secondary ion currents available. Ion detection may be in
the analog mode, and most importantly, conventional mass spectrometry data acquisition systems may be used because of the large ion
currents available. Vacuum requirements are not stringent because
the pressure and residual gas composition are dictated by the
organic liquid being used for the sample matrix.

Discussion

In the previous section, we discussed the general design objectives
of SIMS instruments used for three specific types of analyses: 1)
elemental depth profiling analyses; 2) organic analyses of solid
surfaces; and 3) organic analyses of liquid surfaces. Again, these
design features are shown in Table I. If, however, one examines
Table I more closely, and with a horizontal emphasis instead of the
previous vertical emphasis, one can see that several design features are desirable in all SIMS instruments. Under the **Primary
Beam** category, one sees that all three types of SIMS instruments
should use small diameter bombarding beams. Under **Secondary Ion
Optics**, one will notice that the use of an immersion lens for secondary ion extraction is always desirable, as is an electrostatic
energy analyzer. These two critical categories will be discussed
in this section, but they will be discussed in reverse order,
starting with Secondary Ion Optics, for reasons which will become
apparent.

Secondary Ion Optics

Strictly speaking, the term "secondary ion optics" refers to all
those ion-optical components which transmit secondary ions from the
sample surface to the detector. For the sake of this discussion we
will limit the scope of the term to those secondary ion-optical
components between the sample surface and the mass spectrometer.
With this restriction, we see that the purpose of the secondary ion
optics is to extract the ions efficiently from the sample surface
and present them to the entrance of the mass spectrometer in a
satisfactory manner. Recognition of three important facts are
necessary for such ion optics design:
  1) Secondary ions are emitted from the sample surface
     with a cosine distribution centered about the sample
     surface normal as shown in Figure 2.
  2) The geometrical and/or electrical entrance apertures
     of mass spectrometers are very small.
  3) The Liouville Theorem states that one can extract and
     transmit ions emitted from a large area on the sample
     surface with a narrow angle of divergence, or one can
     extract and transmit ions emitted from the surface
     with a large angle of divergence, but only from a
     small area on the sample surface. But one cannot
     simultaneously extract ions with large angular
     divergence from a large area.

The consequences of these boundary conditions strictly limit the secondary ion transmission efficiency of the instrument and thus, ultimately, the sensitivity of the analysis.

In order to obtain high sensitivity the secondary ion optics must be capable of transmitting to the mass spectrometer those ions which are emitted at large angles with respect to the surface normal. This is best accomplished with an immersion lens(12) placed close to the sample surface as shown in Figure 3. The lens should form a real image of the sample surface in an image plane which then must be "transferred" to the entrance of the mass spectrometer with transfer optics. The net result is that a large fraction of the ions emitted from a point on the sample will be imaged onto the entrance aperture of the mass spectrometer resulting in high transmission efficiency and sensitivity.

At this point, it is appropriate to consider the size of the entrance aperture of the mass spectrometer. Double-focusing magnetic sector instruments have entrance slits only micrometers wide. Clearly, if a real ion-image of the emitting surface is illuminating this aperture, only those ions originating from a very small area of that emitting surface will be passed into the mass spectrometer. Thus, it is inadvisable to design a liquid surface organic SIMS instrument which uses a large diameter (millimeters) bombarding beam because only those ions emitted from a 10-100 micrometer diamater area of the sample surface will be passed into the mass spectrometer. Any primary beam falling outside this area will only contribute to radiation damage of the sample and not to the analytical signal. Therefore, small diameter, focusable ion beams are recommended for this application for most efficient sample utilization.

One might think that quadrupole mass spectrometers would be less susceptible to this limitation because of their much larger entrance apertures. Unfortunately, this is only marginally true, and only for ions of low mass. This is because the electrical entrance aperture of the quadrupole is usually considerably smaller than the physical entrance aperture. By "electrical entrance aperture" we mean that area of the physical entrance aperature through which ions must travel if they are to be transmitted through the quadrupole rod structure. The radius of this aperture, r_q, is small and mass-dependent as seen by the following equation

$$r_q = 2/3 \; r_o \; (\Delta M/M)^{1/2} \qquad (1)$$

where r_o is the radius of the largest circle that can be inscribed within the four quadrupole rods, and $\Delta M/_M$ is the mass resolution of the spectrometer(13).

Consider a quadrupole mass spectrometer with 6mm diameter rods operating at a mass resolving power of 600 (which is needed for high mass organic liquid surface SIMS). The radius of the electrical entrance aperture is only 0.07mm (0.003"). This also makes a very small target for the secondary ion optics to "hit" with the ions being emitted from the sample. The optics can function much more efficiently in transmitting ions from the sample surface into the mass spectrometer if the secondary ions are being

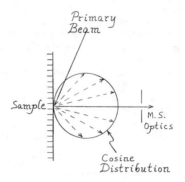

Figure 2. Cosine distributions of sputtered particles. The length
of the arrow is proportional to the probability of ion emission in
that direction.

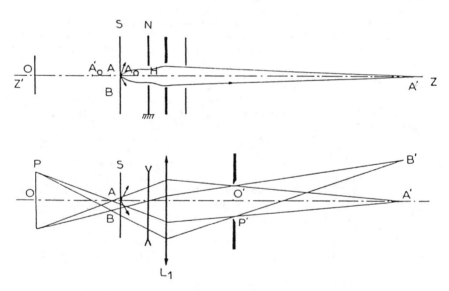

Figure 3. Immersion lens a uniform field E_0 is set up between the
target S and an electrode N (at ground potential). S and N are
parallel. The ions emitted at point A_0 of the flat surface of
object S are accelerated by this uniform field. The ions escape
from the accelerating field through a round hole (T) in N. The
optical axis of the system of collection is the axis (z'z) of the
hole, which lies perpendicular to the surface of the target. The
hole T acts as a diverging lens. A unipotential lens L, with a z'z
axis projects the virtual image A of the surface in a real image
A'. At the same time, this lens produces a real image of the
crossover over which a "contrast" diaphragm can be placed. The
diameter of the diaphragm limits a diameter on the initial virtual
pupil and therefore the sample area from which ions are accepted.
Reproduced with permission from Ref. 5 copyright 1980, Academic
Press.

emitted from a small spot on the sample surface (because of the Liouville theorem). Therefore, small (50-150 um) diameter primary ion beams should always be used for sputtering.

But what about extracting organic secondary ions emitted from a solid surface where one desires to sputter the surface over a large area so that no one spot accumulates a high amount of radiation damage? The Liouville theorem specifically states that one cannot extract a large number of secondary ions (those ions with a large angle of divergence) from such a large area as would be required for a highly sensitive analysis. The solution to this problem of efficient ion extraction from large areas is called "dynamic emittance matching". As first described by Liebl (14), the technique provides a clever ion-optical path around the restrictions of the Louiville theorem.

Optimum transmission of secondary ions is achieved when the acceptance of the mass spectrometer is matched with the secondary ion beam. Figure 4a shows schematically the transfer of the secondary ion beam from the sample to the mass spectrometer entrance slit. If d_1 is the diameter of the area to be analyzed, the entrance slit (width s) has to accept the image $d_1 + \delta_1$ which is $d_2 + \delta_2$. δ_1 is the virtual width of the sample point, given by $\delta_1 = V_i/E$ where V_i is the initial kinetic energy and E is the field strength on the surface. This explains the desirability for high extraction fields at the sample surface). Using the Louiville theorem, we obtain

$$(d_1 + \delta_1) \, \alpha_1 = s \, \alpha_2 \qquad (2)$$

This results in a certain beam divergence α_2 of which only the angle α is accepted by the mass spectrometer. This is the case whether a large, stationary beam irradiates the entire width d_1, or a fine beam probe is raster scanned over d_1. This represents the state of the art for virtually all solid-surface organic SIMS instruments. However, if a fine scanned primary beam is used, the transmission can be improved considerably by using dynamic emittance matching shown in Figure 4b. In this scheme, the secondary ion beam transfer optics incorporates a deflector synchronized with the primary beam movement so that the movement of the virtual source spot δ_1 is cancelled. Instead of the previous equation, the following condition must now be met

$$d_1 \alpha_1 = s \, \alpha_2 \qquad (3)$$

with the same slit width S and α_1. The final beam divergence α_2 in this case is now smaller by a factor of $(d_1 / \delta_1) + 1$ and accordingly, more of the secondary ions fall within the acceptance angle of the mass spectometer.

Dynamic emittance matching has been applied with great success in the field of organic solid surface SIMS analysis by Campana and co-workers (15) and by Benninghoven and co-workers (16), but due to the sophistication required, this technique has not yet seen wide use in the field. Nevertheless, as analytical requirements call for higher sensitivity, especially for high mass ions, dynamic emittance matching will be required.

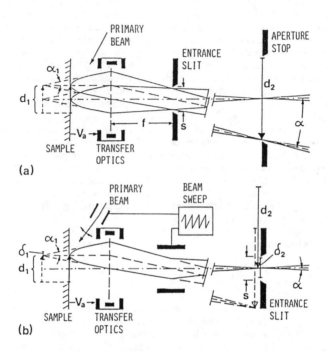

Figure 4.   Secondary ion Transfer from sample to mass spectrometer.

a) static beam transfer
b) dynamic beam transfer (dynamic emittance
   matching)

Future Trends in SIMS Instrumentation

While secondary ion mass spectrometry has been practiced for over
30 years, the technique is far from "mature". In fact, in the organic SIMS field, the technique is still in its infancy. Most of
the desirable instrumental features outlined in this paper for organic SIMS instruments have yet to be incorporated into present-day
machines. Most instruments currently being used for organic SIMS
were originally designed for quite different purposes and have
capabilities for SIMS far below those which are potentially available if currently available ion-optical techniques are used. This
is particularly true in the liquid-surface organic SIMS field.
None of the instruments use finely focussed ion beams to sputter
only that area on the sample surface which will produce ions that
can be transmitted by the mass spectrometer. For reasons which are
unclear, the makers of magnetic sector organic SIMS instruments
promulgate the mystique that it is difficult to direct a focussed
ion beam into the sample region due to the high electric fields
present when, in fact, this has been done ever since the beginning
of SIMS (17).
        Secondary ion extraction optics are also far from optimized.
No currently commercially available organic SIMS instruments
utilize an immersion lens for ion extraction. There is no question
that the instruments in current use, even though not optimally
designed for this purpose, have provided a wealth of information
for organic chemists and mass spectrometrists. But what instrumentation are we going to need to solve tomorrow's problems? An
area that is becoming increasingly important is the molecular
characterization of solid surfaces. At present, we have no
analytical techniques to probe this type of sample other than
organic SIMS. Elemental characterization of solid surfaces can be
obtained using a variety of techniques (SIMS, Auger electron
spectrometry, X-ray photoelectron spectrometry, etc.). But when it
comes to answering the question, "What molecules are present on my
sample surface?", we are in great need of more sophisticated SIMS
instrumentation. Besides an instrument with the features described
here (small ion beams used with dynamic emittance matching, high
transmission extraction optics, etc.), there are several other
instrumental approaches which could provide exciting possibilities
for high sensitivity molecular surface mass spectrometry. Optimization of time-of-flight instrumentation has obvious advantages due
to high ion transmission and detection of all transmitted species.
This area will undoubtedly see a great increase in interest in the
near future.
        Another exciting possibility for high sensitivity molecular
surface mass spectrometry is the use of laser-excited ion desorption in a pulsed ion cyclotron resonance experiment using Fourier
transform techniques. In an ideal situation, this scheme could include all those attributes which are desirable for solid-surface
molecular characterization:
    1) surface sensitivity due to laser desorption
    2) high "transmission" due to the high collection efficiency
        of the pulsed ICR cell approach
    3) high mass resolution
    4) high mass range

These areas should provide some exciting avenues to explore in the coming years for organic SIMS.

But what about inorganic SIMS? Here the field is indeed more mature with the advent of the ims-3f by CAMECA:Thomson-CSF in 1978 (5). But there are still a few interesting instrumental aspects to explore, one of which is the application of ultra-finely focussed ion beams (18).

In the last several years, researchers have developed the liquid-metal field-ionization source to a high level of performance and reliability. Developed mainly for direct-writing ion implantation and ion beam lithography, these sources have the potential of greatly enhancing the capabilities of inorganic depth-profiling SIMS instruments. Their high brightness makes it possible to form beams <1000Å in diameter with current densities of many A/cm^2. Unfortunately, however, the beam species available are limited mainly to Ga$^+$ and In$^+$ from metals which have low melting points and low chemical reactivity. These species produce none of the yield-enhancing characteristics (19) needed for high sensitivity analyses and may even reduce the yield of positively charged ions. In addition, these ion sources, due to their exceedingly small beam size not withstanding their high current densities, are severely limited in the total ion current deliverable to the target. This means that in order to obtain high sensitivity with these small beams, only the highest transmission mass spectrometers should be used. This limitation essentially rules out all quadrupole type instruments due to their limited transmission of ions with a wide energy spread as is the case with atomic secondary ions.

The liquid metal ion source could help dramatically in the area of negative secondary ion mass spectrometry where bombarding with Cs$^+$ ions has proven very beneficial (19). This is due to the large increase in negative ion emission from a cesiated, low work function surface. Unfortunately, the sources used to produce cesium ions are of the surface ionization type (20) which has a low brightness. If the liquid metal field emission ion source can be developed to operate reliably with liquid cesium, then small, high-current density ion beams could be formed. Currently, however, the reactive nature of the metal makes the liquid metal field emission cesium ion source more of a research project than a routine analytical technique.

Summary

It has been the purpose of this paper to provide an overview of the basic differences and similarities of the various types of instruments which detect ionized particles emitted from surfaces by energetic particle bombardment. Since the scope of secondary ion mass spectrometry is so broad, it is not surprising that no one instrument has been designed to perform optimally for all types of SIMS analyses. Design aspects of the primary beam, extraction optics, mass spectrometer, detection equipment and vacuum system must be considered to construct an instrument best suited for a particular purpose.

Acknowledgments

The author gratefully acknowledges the many helpful discussions with P. J. Gale and B. L. Bentz concerning this work, as well as the help of W. L. Harrington and R. E. Honig concerning the manuscript.

Literature Cited

(1) C. W. Magee and R. E. Honig, Surf. Interface Analysis 4, 35 (1982).

(2) C. W. Magee, W. L. Harrington and R. E. Honig, Rev. Sci. Instrum. 49, 477 (1978).

(3) J. A. McHugh in "Methods of Surface Analysis", A. W. Czanderna, Ed.; Elsevier, Amsterdam, 1975, pp.223-278.

(4) K. Wittmaack and J. B. Clegg, Appl. Phys. Lett. 37, 285 (1980).

(5) G. Slodzian in "Advances in Electronics and Electron Physics, Supplement 13B", Academic Press, N.Y. 1980, pp.1-44.

(6) B. N. Colby and C. A. Evans, Jr., Appl. Spectroscopy 27, 274 (1973).

(7) E.g., Princeton Applied Research: Model 1120.

(8) A. Benninghoven, D. Jaspers and W. Sichtermann, Appl. Phys. 11, 35 (1976).

(9) C. W. Magee, Int. J. Mass Spectrom. Ion Phys. 49, 211 (1983).

(10) M. Barber, R. S. Bardoli, G. J. Elliot, R. D. Sedgwick and A. N. Tyler, Anal. Chem. 54, 645A (1982).

(11) P. Sigmund in "Sputtering by Ion Bombardment - I.", R. Behrisch Ed.; Topics in Applied Physics, Springer-Verlag, Berlin, 1981, pp.9-71.

(12) For a discussion on various types of immersion lenses see: K. J. Hanszen and R. Lauer in "Focussing of Charged Particles"; A. Septier, Ed.; Academic Press, N.Y., (1978), pp.296-300 .

(13) P. H. Dawson, "Quadrupole Mass Spectrometry and Its Applications", Elsevier, Amsterdam, 1976.

(14) H. Liebl in "Advances in Mass Spectrometry, Vol. 7A", N. R. Daly (Ed.), Heyden and Son, London, 1978, pp.751-757. Also H. Liebl in "Low Energy Ion Beams, 197), Conf. Series No. 38", K. G. Stephens, I. H. Wilson and J. L. Moruzzi, Eds.; Institute of Physics, Bristol, London, 1978, pp.266-281. Also Liebl, Nucl. Instrum. Methods 187, 143 (1981).

(15) J. E. Campana, J. J. DeCorpo and J. W. Wyatt, Rev. Sci.
     Instrum. 52, 1517 (1981).

(16) R. Jede, O. Ganschow and A. Benninghoven in "Secondary Ion
     Mass Spectrometry - SIMS III", A. Benninghoven, J. Giber,
     J. László, M. Riedel and H. W. Werner, Eds.; Series in
     Chemical Physics, Vol. 19, Springer-Verlag, Berlin, 1982,
     PP.66-71.

(17) R.F.K. Herzog and F. P. Vieböck, Phys. Rev. 76,
     855L-856L,(1949).

(18) For a collection of many papers on field emission ion sources,
     see The Proceedings of 16th Annual Symposium on Electron, Ion
     and Photo Beam Technology, J.V.S.T. 19, 1145-1190 (1981).

(19) H. A. Storms, K. F. Brown and J. D. Stein, Anal. Chem. 49,
     2023 (1977).

(20) For a discussion of surface ionization sources, see:
     R.G. Wilson and G. R. Brewer, "Ion Beams", John Wiley and Son,
     N. Y. (1933), pp.26-32, 72-77. Surface ionization Cs^+ sources
     are commercially available for General Ionex Corp., Newbury-
     port, MA.

RECEIVED April 16, 1985

Liquid Metal Ion Sources

Douglas F. Barofsky

Department of Agricultural Chemistry, Oregon State University, Corvallis, OR 97331

Use of liquid metal ion (LMI) sources to produce
secondary ions in the dynamic SIMS mode has yielded
ion abundances of sugars up to 50 times greater than
those produced with neutral argon beams and in the
static SIMS mode has generated useful molecular
weight information from as little as 6 x 10^{-17} mole
of crystal violet and 2 x 10^{-13} mole of leucine-
enkephalin. LMI sources permit broad variation of
the mass and molecular type of the bombarding
particles, a potentially useful feature for studies
of secondary ionization of organic molecules.
Construction of LMI sources, attachment to existing
mass spectrometers, and operation are relatively
simple.

The liquid metal ion (LMI) source is an electrohydrodynamic field
emitter. Experimentally, it is known that emission takes place from
a cone of liquid metal (1) which is formed by application of a
relatively high electric potential and maintained by the resulting
high electric field (2,3). The ion source operates with a variety
of metals. For any given metal both ions and charged microparticles
are emitted. The ionic component is comprised of singly and
multiply charged atomic and molecular ions; Table I lists some of
the ion species produced by LMI sources. The relative abundances of
these different ions as well as their individual energy
distributions and angular divergences are strongly dependent on the
total emission current (4-15). Although it has not been studied
as thoroughly as the ionic part, the charged microparticle
component apparently accounts for about two-thirds of the ejected
material at emitter currents less than 100 μA (16), and both the
size and size range of the charged microparticles have been shown
to increase measurably with increasing emission current (17,18).
At present there are many unknown aspects concerning the ionization
and ejection processes occurring at the apex of the emission cone;
hence, a theoretical model consistent with the LMI source's
emission properties has not yet been formulated.

0097-6156/85/0291-0113$06.00/0
© 1985 American Chemical Society

Table I. Selection of Ions Produced by LMI Sources

	Ion Species (m/z, most abundant isotope)	
Element / Z	+2	+1
Al	13.5	27, 54, 81, 108, 135
Si	14	28, 56
Ni	29	58, 116
Ga	34.5	69, 138, 207, 276
Se		80
Ag		107
In	57.5	115, 230
Sn	60, 180, 300, 420 540	120, 240, 360, 480 600, 720, 840
Cs		133
Au	98.5, 295.5	197, 394, 591
Pb	104, 312	208, 416, 624
Bi	104.5, 313.5, 522.5	209, 418, 627
U		238

A LMI emitter is virtually a point source; thus, its principal, distinguishing, optical characteristic is its very high brightness which can be shown for a typical emitter to be on the order of 10^6 A/cm^2 sr (19). Focused beams that deliver 20–120 pA into spots of about 100 nm diameter at a convergence half-angle of 1 mrad (1–5 x10^5 A/cm^2 sr) have been achieved recently in the development of LMI sources for ion microprobe analysis of solids (20,21). This brightness feature is important to several charged-particle-beam techniques in microfabrication, ion beam lithography, maskless ion implantation, ion microprobe analysis, and ion propulsion (2,3,22,23). Consequently, there is considerable scientific and engineering interest in LMI sources.

Recently, this investigator and his colleagues began using LMI sources to produce ions from liquid organic solutions (24-28). This application was motivated by the potential for exploiting the LMI source's rich selection of ions and microparticles to investigate secondary ionization processes, the possibility of improving sensitivity, and the more remote prospect of achieving localized, molecular analysis of biological tissues with a finely focused ion beam.

The construction and operation of LMI sources are briefly described in the first section of this paper. The potential use of LMI sources in fundamental investigations of secondary ion emission from liquid organic matrices is discussed in the following section. The few existing preliminary reports on LMI/SIMS combinations used or studied for analytical purposes are reviewed in the final section.

Fabrication and operation

The two common configurations for LMI emitters are the capillary (2) and the needle (3), Figure 1. As its name suggests the

former, Figure 1a, utilizes a narrow capillary. Molten metal is
drawn through the capillary from a reservoir to the end where the
cone of liquid metal is formed. Although they can have very long
lifetimes, these emitters are difficult to make and, to the best of
this writer's knowledge, have not yet been employed for SIMS of
organic compounds.

A needle source consists of a hairpin filament ($\sim$180 µm
diameter), usually of a refractory metal such as tungsten, with a
short length of smaller diameter ($\sim$125 µm) wire spot-welded to it,
Figure 1b. The tip of the latter wire, the emitter, is
electrochemically etched to a point with a radius of curvature at
the apex of 2-5 µm; the etching technique for tungsten has been
described in detail by others (7,29). As quickly as possible
after the assembly is thermally cleaned under vacuum ($\sim$10^{-4} Pa),
the emitter is dipped into a molten pool of liquid metal and then
withdrawn. If done correctly, the junction formed by the bend in
the filament and the emitter wire will hold a small bead of metal,
and the emitter will appear shiny from the thin film of metal on
its surface.

Certain details are important to the successful wetting and
operation of a LMI source. The emitter wire must be of an element
or alloy which is wettable by the liquid metal to be used in the
source. Thermodynamical concepts and data found in handbooks (30)
or reference books (31,32) can be useful for selecting compatible
emitter-metal/liquid-metal combinations. It is generally necessary
to remove the emitter's metal oxide skin before it can be wetted;
this is accomplished in the electrochemical and thermal cleaning
steps. Wetting of the emitter's shank and sharpened tip is further
facilitated by chemical etching to groove and roughen their
surfaces (7,29). The metal in the wetting reservoir should be
kept scrupulously free of dust, oil, or other dirt. Scum on the
reservoir's surface will be transferred to the emitter during
wetting, and the source will not operate correctly or at all. It is
useful to provide the wetting chamber with a viewing port to
observe the emission pattern of a freshly wetted emitter as a
simple performance test before removing it from the vacuum.

A LMI emitter's smallness makes it possible to attach it to
the ion source of almost any magnetic sector or quadrupole mass
spectrometer. Ion sources which have already been designed for fast
atom bombardment (FAB) or field desorption (FD) are ideally suited
to modification for LMI/SIMS operation.

For most analytical applications the LMI source can be
employed in a simple unfocused form. A diagram of an unfocused LMI
source appendaged to the side of a magnetic sector instrument's ion
source is shown in Figure 2. Figure 3 exhibits a photograph of an
actual source of this same design. More sophisticated LMI sources,
which can be focused (24,28,33) and/or pulsed (33), have been
designed to investigate ionization mechanisms and test sensitivity
limits.

LMI sources with needle emitters operate in essentially the
same way as field ionization or field desorption sources. The
filament is resistively heated to melt the metal film and/or
promote its flow to the tip of the emitter. Typically, the emitter
or anode is positively biased 3-5 kV with respect to its counter
electrode, the cathode; the actual operating voltage is determined

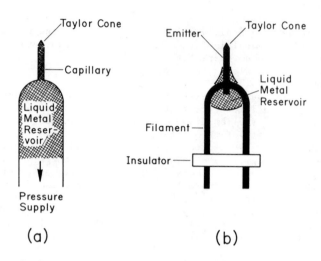

Figure 1. Liquid metal ion emitters: (a) capillary, (b) needle.

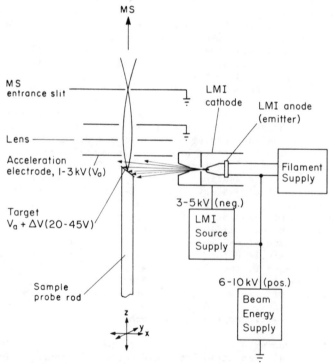

Figure 2. Schematic of a nonfocusing LMI source attached to the ion source of a single-focusing magnetic-sector mass spectrometer.

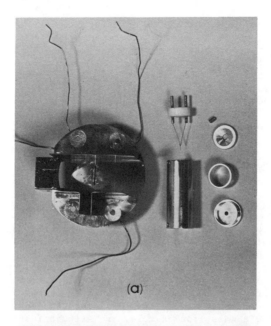

Figure 3. Photograph of a nonfocusing LMI source: (a) unassembled, (b) assembled and attached to the ion source of a single-focusing magnetic-sector mass spectrometer.

by the emission onset voltage, a quantity which is geometry
dependent, and the desired emission current. The anode current is a
convenient parameter for monitoring the source's emission. In an
unfocused configuration the cathode can be maintained at the same
potential as the acceleration electrode in the mass spectrometer
ion source. This arrangement has the disadvantage, however, that
the impact energy of the primary ions and the emission current are
both determined by the LMI source's operating voltage. In order to
separately control the primary beam current and energy, it is
necessary to be able to vary the anode-cathode potential, i.e.the
LMI source's operating voltage, independently of the emitter's
potential. One method for accomplishing this is to float the power
supply for the LMI source as illustrated schematically in Figure 2.

Investigation of secondary ionization processes

One of the most significant uses of LMI sources in connection with
SIMS of organic compounds may be as probes in performing
measurements of secondary particle yields. Such measurements are
important for understanding the processes of secondary ion emission
from solid and liquid organic samples. Total particle yields
reflect directly the dynamical aspects of emission processes;
variations in primary beam energy, incident flux density, and
primary particle mass, for example, are all manifested in changes
in total particle yields. The ratio of secondary ion yield to total
particle yield and the ratio of secondary ion yields from two
different species can be sensitive, quantitative monitors of the
chemistry and kinetics, respectively, of ionization processes.

The particular advantage LMI sources offer to such experiments
is a selection of prospective primary particles that spans a broad
range of mass and molecular type. Some of the ions available from
LMI sources are listed in Table I. Among the particles tabulated
there are several groupings within which the effects of various
particle types can be probed essentially independently of particle
mass, e.g. $^{29}Si_2^+$ (58), $^{58}Ni^+$ (58), $^{116}Sn^{++}$ (58); $^{107}Ag^+$ (107),
$^{27}Al_4^+$ (108); $^{27}Al_5^+$ (135), $^{69}Ga_2^+$ (138); $^{69}Ga_3^+$ (207), $^{209}Bi^+$ (209);
$^{197}Au_3^{++}$ (295.5), $^{118}Sn_5^{++}$ (295); and $^{207}Pb_2^+$ (414), $^{118}Sn_7^{++}$ (413).

For example, experiments might be conducted with ions from one or
more of these groups to distinguish between linear and nonlinear
collision processes (34). This author and his co-workers recently
described an apparatus that combines a LMI source and a Wien filter
to produce a focused beam of particles which are selected by mass
and type (28). When completed, this versatile probe will be used
to investigate secondary emission from liquid organic matrices.

This same group of investigators has also been developing a
combined gravimetric/scintillation counting technique for measuring
total secondary particle yields from liquid organic solutions
(28,35). Preliminary results, Figures 4 and 5, from measurements
on glycerol solutions are surprising. The total particle yields
from glycerol bombarded by 3-5 keV indium ions are in the range of
1000-4000, two to three orders of magnitude higher than those
generally observed for metals in this energy range. The yields for
moderately concentrated sugar solutions appear to be twice as large

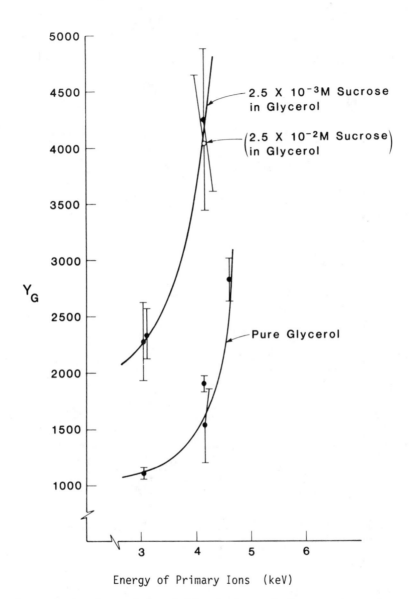

Figure 4. Total particle yields of glycerol versus primary ion energy: In-LMI source.

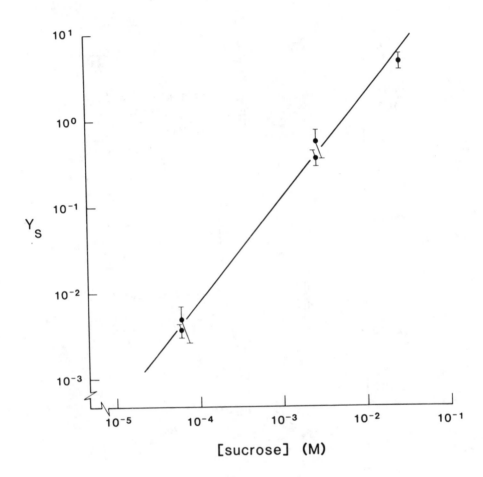

Figure 5. Total particle yields of sucrose versus sucrose concentration in glycerol: In-LMI source.

as those for pure glycerol. The yields for both the pure material
and the solutions are nonlinearly increasing functions of energy.
Sucrose yield as a function of solution concentration, Figure 5, is
essentially linear over the range studied. Positive and negative
secondary ionization efficiencies of sucrose were estimated from
cursory ion yield measurements to be about 1×10^{-5} and 4×10^{-4}
respectively. Results similar to those for sucrose were obtained
for the $(M+1)^+$ ion of adenosine.

Analytical applications

This author and his associates initiated their work with LMI
emitters to determine if they would be useful in fundamental
investigations of secondary ion emission from liquid organic
matrices. As a result, the initial experiments with LMI sources
were performed on samples dissolved in glycerol in the same manner
as is commonly done for FAB. Unusually high signals for the
molecular ions of the sample compounds were immediately noted in
these experiments ($\underline{24,25,27}$). For example, the abundance of the
$(M + Na)^+$ ion of stachyose produced from a glycerol matrix by
bombardment with gold ions was found to be about 50 times that
produced with a neutral argon beam and about 15 times that produced
with a neutral xenon beam; the energies and current densities at
the target were comparable for all of the primary beams.

The large gain in molecular ion abundance obtained by changing
from a primary beam of argon atoms to one of gold ions cannot be
explained on strictly theoretical grounds and may be partially due
to differences in mounting the primary sources ($\underline{27}$). Regardless
of its explanation, it demonstrates that alternate primary source
conditions exist which can significantly increase the sensitivity
in SIMS of compounds in liquid organic solutions. Examples which
demonstrate the potential utility of this increased sensitivity
have been published elsewhere ($\underline{24,25,27}$).

Because of its point source characteristic, the LMI emitter is
particularly well suited to producing finely focused beams. It may
be possible to take advantage of this focusing characteristic to
increase the sensitivity of SIMS for organic compounds. Focusing
the primary beam to a very small spot on the sample target can,
with the proper source geometry, reduce the area which must be
covered by sample. This could be taken advantage of to
significantly reduce the actual amount of sample required for mass
analysis and, consequently, achieve an increase in analytical (not
just instrumental) sensitivity. Because neutral atom beams cannot
be focused, it would not be possible to implement this method of
increasing sensitivity with FAB sources. In this respect an ion
beam is superior to an atom beam.

Chiat and Field ($\underline{33}$) have recently investigated the use of a
focused, pulsed LMI source in combination with a time-of-flight
mass spectrometer for increasing the sensitivity of analyses by
static SIMS. Their preliminary results are noteworthy. Using a
primary beam focused to 100 μm diameter and average ion current
densities of $0.6-60$ nA/cm^2 at the target, they were able to obtain
useful molecular weight information from as little as 1×10^{-15} mole
of crystal violet and 3×10^{-12} mole of leucine-enkephalin, a
neuropentapeptide. Actually, only 6% of the total sample introduced

into the mass spectrometer was bombarded by the primary beam indicating that the mass spectral information was being obtained on just 6×10^{-17} mole and 2×10^{-13} mole respectively of these two compounds. These investigators concluded from there findings that substantial improvement in sensitivity can be obtained by applying sample only to the area struck by the primary beam and that the inherent limitation in sensitivity by this method will arise in practice from an inability to handle very small amounts of material cleanly.

Stoll, et al (36) recently reported the operation of an ion microprobe that indirectly demonstrated the potential for higher sensitivity in the dynamic SIMS mode when a finely focused primary ion beam is employed. The ion microprobe consisted of a focused LMI source, capable of producing a beam spot $\leq$ 1 μm and an ion current density on the order of 1 A/cm² at the target, attached to a high resolution double focusing mass spectrometer (VG ZAB-2F) (37). To maintain both optimum acceptance of the mass spectrometer and spatial resolution of the primary ion beam over a relatively large field of view (6 mm x 6 mm), two, high-precision, piezoelectric linear motors were used to move the sample probe in a X-Y plane perpendicular to the incident primary ion beam without deflecting the latter. This system was used to image a fine Ni-wire mesh (500 mesh, 25 μm wire diameter) with secondary ions of various organic compounds which had been deposited on the mesh; for example, a molecular image of the mesh was produced with a Fomblin ion at 1281 Daltons. Because of the very small beam dimension, the molecular weight of the imaging ion, and the use of a magnetic sector instrument, this experiment indirectly indicates a high sensitivity. It suggests that only a very small absolute amount of liquid sample solution would be required for analysis with this instrument.

Although only limited results have been reported to date, LMI sources do appear to have potential for practical, analytical applications in SIMS of organic compounds.

Literature cited

1. Taylor, G. I.   Proc. R. Soc. London, Ser. A 1964, 280, 383-97.
2. Mahoney, J. F.; Yahiku, A. Y.; Daley, H. L.; Moore, R. D.; Perel, J.   J. Appl. Phys. 1969, 40, 5101-6.
3. Clampitt, R.; Aitken, K. L.; Jefferies, D. K.   J. Vac. Sci. Technol. 1975, 12, 1208.
4. Sakurai, T.; Culbertson, R. J.; Robertson, G. H.   Appl. Phys. Lett. 1979, 34, 11-13.
5. Sudraud, P.; Colliex, C.; van de Walle, J.   J. Phys. (Orsay, Fr.) 1979, 40, L207-11.
6. Mair, G. L. R.; von Engel, A.   J. Appl. Phys. 1979, 50, 5592-5.
7. Swanson, L. W.; Schwind, G. A.; Bell, A. E.; Brady, J. E.   J. Vac. Sci. Technol. 1979, 16, 1864-7.
8. Culberton, R. J.; Robertson, G. H.; Sakurai, T.   J. Vac. Sci. Technol. 1979, 16, 1868-70.

9. Swanson, L. W.; Bell, A. E.; Schwind, G. A.; Orloff, J.
   Proc. Sympos. Electron Ion Beam Sci. Technol., 1980, p. 594.
10. Swanson, L. W.; Schwind, G. A.; Bell, A. E. J. Appl. Phys.
    1980, 51, 3453-5.
11. Gamo, K.; Ukegawa, T.; Namba, S. Jpn. J. Appl. Phys. 1980,
    19, L379-82.
12. Gamo, K.; Ukegawa, T.; Inomoto, Y.; Ka, K. K.; Namba, S.
    Jpn. J. Appl. Phys. 1980, 19, L595-8.
13. Dixon, A.; Colliex, C.; Sudraud, P.; van de Walle, J. Surf.
    Sci. 1981, 108, L424-8.
14. Bell, A. E.; Schwind, G. A.; Swanson, L. W. J. Appl. Phys.
    1982, 53, 4602-5.
15. Ishitani, T.; Umemura, K.; Hosoki, S.; Takayama, S.; Tamura, H.
    J. Vac. Sci. Technol. A 1984, 2, 1365-9.
16. Mair, G. L. R.; von Engel, A. J. Phys. D 1981, 14, 1721-27.
17. Thompson, S. P.; von Engel, A. J. Phys. D 1982, 15, 925-31.
18. D'Cruz, C.; Pourrezani, K.; Wagner, A. Proceedings of the
    1984 International Symposium on Electron, Ion, and Photon
    Beams, 1984, p. 3.
19. Komuro, M.; Hiroshima, H.; Tanoue, H.; Kanayama, T. J. Vac.
    Sci. Technol. B 1983, 1, 985-9.
20. Seliger, R. L.; Ward, J. W.; Wang, V.; Kubena, R. L. Appl.
    Phys. Lett. 1979, 34, 310-12.
21. Ishitani, T.; Tamura, H.; Todokoro, H. J. Vac. Sci. Technol.
    1982, 20, 80-3.
22. Krohn, V. E.; Ringo, G. R. Appl. Phys. Lett. 1975, 27,
    479-81.
23. Clampitt, R.; Jefferies, D. K. Nucl. Instrum. Methods 1978,
    149, 739-42.
24. Barofsky, D. F.; Giessmann, U.; Swanson, L. W.; Bell, A. E.
    Proc. 29th Int. Field Emission Sympos., 1982, pp. 425-32.
25. Barofsky, D. F.; Giessmann, U.; Swanson, L. W.; Bell, A. E.
    Int. J. Mass Spectrom. Ion Phys. 1983, 46, 495-7.
26. Giessmann, U.; Barofsky, D. F. 31st Ann. Conf. Mass
    Spectrom. Allied Topics, 1983, pp. 602-3.
27. Barofsky, D. F.; Giessmann, U; Bell, A. E.; Swanson, L. W.
    Anal. Chem. 1983, 55, 1318-23.
28. Barofsky, D. F.; Murphy, J. H.; Ilias, A. M.; Barofsky, E. In
    "Secondary Ion Mass Spectrometry SIMS IV"; Benninghoven, A.;
    Okano, J.; Shimizu, R.; Werner, H. W., Eds.; SPRINGER SERIES IN
    CHEMICAL PHYSICS 36, Springer: Berlin, 1984, pp. 377-9.
29. Wagner, A.; Hall, T. M. J. Vac. Sci. Technol. 1979, 16,
    1871-4.
30. Glang, R. In "Handbook of Thin Film Technology"; Maissel, L.
    I.; Glang, R., Eds.; McGraw-Hill: New York, 1970; Chap. 1.
31. Holland, L. "Vacuum Deposition of Thin Films"; John Wiley &
    Sons: New York, 1956; Chaps. 3,7.
32. Loonam, A. C. In "Vapor Deposition"; Powell, C. F.; Oxley, J.
    H.; Blocher, Jr., J. M., Eds.; John Wiley & Sons: New York,
    1966; Chap. 2.
33. Chait, B. T.; Field, F. H. 32nd Ann. Conf. Mass Spectrom.
    Allied Topics, 1984, pp. 237-8.
34. Sigmund, P. In "Inelastic Ion-Surface Collisions"; Tolk N. H.;
    Tully, J. C.; Heiland, W.; White, C. W., Eds.; Academic Press:
    New York, 1977; pp. 121-52.

35. Barofsky, D. F.; Ilias, A. M.; Barofsky, E.; Murphy, J. H. 32nd Ann. Conf. Mass Spectrom. Allied Topics., 1984, pp. 182-3.
36. Stoll, R. G.; Harvan, D. J.; Cole, R. B.; Hass, J. R. 32nd Ann. Conf. Mass Spectrom. Allied Topics., 1984, pp. 838-9.
37. Stoll, R. G.; Harvan, D. J.; Hass, J. R.; Giessmann, U.; Barofsky, D. F. 31st Ann. Conf. Mass Spectrom. Allied Topics., 1983, p. 123.

RECEIVED April 16, 1985

8

Fast Atom Bombardment Mass Spectrometric Technique and Ion Guns

Julius Perel

Phrasor Scientific, Inc., Duarte, CA 91010

The FABMS (Fast Atom Bombardment Mass Spectrometry)
technique is reviewed in this chapter with emphasis
upon laboratory instrumentation and procedures.
Several popularly used primary beam guns are described
including the effects of the bombarding species.
Illustrative equations relating secondary ion currents
to the primary impingement rates are presented. Geometrical arrangements between the primary beam, the
sample target surface and the analyzer axis are illustrated. Solutions are suggested to overcoming limitations encountered in retrofitting FAB on mass spectrometers. Testing procedures are outlined along with
matrix considerations. Sources of noise generated
during particle impingement are examined. Some new
experimental techniques are discussed which lead to
increasing signal-to-noise ratios. Finally, SIMS
(Secondary Ion Mass Spectrometry) is shown to be
available to spectrometrists who have FABMS.

Background

Although the FABMS (Fast Atom Bombardment Mass Spectrometry) technique has only been in use for a few short years (since 1981), it
traces its roots back well over a century (1). It has been observed
that the bombardment of a surface by energetic ions produces the desorption of atoms and molecules from the surface of the target. This
process, known as Sputtering, produces a yield (number of atoms sputtered per incident ion) which generally increases with the energy,
the mass and the incidence angle of bombardment (2).
   Only a small fraction, perhaps 1 percent of the sputtered particles, were found to be charged (3). Both positive and negative
charges were found, depending upon the target materials. When mass
analysis was applied to the process, it was then known as Secondary
Ion Mass Spectrometry (SIMS). Subsequently, SIMS has been an important aid in investigating the surface material, the underlying composition can also be analyzed, leading to the technique of depth profiling. Methods were developed to deposit organic samples on

0097-6156/85/0291-0125$06.00/0
© 1985 American Chemical Society

surfaces which were then subjected to SIMS. This was known as
Organic SIMS or Molecular SIMS, and became interesting to bio-
chemists (3-5). Non-volatile organic samples can be analysed using
this technique. However, this method suffers from short sample lifetime. When a sample exists as a monolayer, one is limited to about
10^{12} molecules on a typical spectrometer target area. Thus low bombarding ion current densities were required to extend the lifetime,
but this in turn resulted in low signals (1, 6).
     An innovation introduced by Barber and colleagues (7-9) allowed
the observation of non-volatile and thermally labile molecules with
adequate sample lifetime. The sample compound, dissolved in glycerol,
was placed on a probe tip and introduced into the spectrometer. Bombardment by fast neutral atoms produced desorbed molecular ions with
some fragmentation spectra. Fast atoms were chosen because they were
not slowed or deflected by the positive voltage on the ionization
source of the magnetic sector mass spectrometer used on these first
tests and they avoided charging of the sample target. Although this
technique became known as Fast Atom Bombardment Mass Spectrometry
(FABMS), the new innovation was the use of the liquid matrix to contain the sample. Well known in the earlier sputtering and SIMS research, charging was not a problem and fast ions perform equally well
as fast atoms (1, 10, 11). Other names, such as Liquid SIMS have
been suggested for this process, but so far FABMS, a misnomer, remains rooted. FABMS should be defined as a process of atom and/or
ion bombardment of a liquid containing a sample compound dissolved or
floating in order to desorb molecular ions for analysis with a mass
spectrometer.

Guns That Produce the Primary Beams

Several guns have been used to produce the primary beam to bombard
the FABMS target (12). These beams are composed of fast neutral
atoms or ions, or a mixture of the two. It does not appear that the
charge state is critical, because desorption is produced by momentum
transfer. Some of the guns and the characteristics of the beams are
examined in this section.

The Saddle-Field Gun   The Saddle-Field Gun, capable of producing both
fast atoms and ions, was first commercially introduced for FABMS
analysis (7, 13). Electron oscillations are induced in the gun chamber by electrostatic fields applied between the positive anode and
the negative (grounded) cathode as seen in Figure 1. Electrons
accelerated from the cathode region toward the anode, pass through
the central openings because of the field configuration. Upon
approaching the cathode on the other side, they are slowed, then re-
turn toward the anode again to repeat the reverse trajectory. Ions
are generated by electron impact upon the gas atoms introduced into
the gun chamber, thus forming a plasma in which new electrons are
generated. Ions accelerated toward the upper anode can pass through
the aperture to form an ion beam. The beam current is controlled by
varying the voltage and/or the gas flow. Fast atoms are believed to
arise when the fast ions capture an electron during an ion-electron
collision (13). Because of the more favorable collision cross sections, it appears more probable that fast atoms arise from ion-atom
charge transfer collisions.

The Capillaritron The Capillaritron ion source (Figure 2) consists
of a fine bore (25μm) capillary nozzle and a concentric extractor
electrode (14-16). By flowing a gas or vapor through the capillary,
a microdischarge or plasma is formed at the exit orifice when a voltage (> 2kV) is applied between the capillary and the grounded extractor electrode. The small dimensions of the capillary orifice and
overall nozzle shape are important factors in the initiation and
stability of the plasma and subsequent ion/atom beam. These small
dimensions allow guns to be designed to accommodate tight areas such
as in the case of the DIP Gun. Ions generated by the plasma are
accelerated by the electric field at the nozzle tip to form a beam.
Some of the ions, when colliding with gas atoms emerging from the
orifice, are neutralized by charge transfer and so become fast atoms.
These fast atoms compose approximately half of the energetic beam
(15). The beam current is also controlled by the voltage and gas
flow rate. Typically xenon is recommended as the gas feed into the
gun because of its high mass, but argon or virtually any other gas
can produce FAB spectra.

The Cesium Ion Gun The Cesium Gun was introduced as a FABMS gun
because the ion mass is large, close to that of xenon, and because it
does not produce a gas load (17). Gas fed guns produce a gas load
which degrades spectra on spectrometers having insufficent pumping
speed. Operation of this gun can be explained with the help of
Figure 3. Ions are thermionically emitted from the alkali alumino-
silicate solid maintained at a high temperature (~ 1000°C) where the
alkali species have relatively free movement. Upon desorption from
the surface the ions are accelerated by the electric fields applied
between the emitter and the accel or grounded electrode. Total emission current from the heated emitter surface is a strong function of
the temperature. Virtually no fast or slow atoms are present and the
ion beam energy distribution is very narrow. This narrow energy
spread with the help of the lens shown in the figure, allows for beam
focusing which is not readily available with some of the gas sources.
The Cesium Gun has been reported to produce cleaner spectra with
lower noise than the gas-type gun (18).

Other Primary Beam Guns The Liquid Metal Ionization (LMI) source,
also a non-gas-fed source, operates on the principle of producing a
cone-like apex on the surface of a liquid metal exposed to an intense
electric field. At the liquid cone apex, the fields are sufficiently
high (> 10^9V/m) to extract ions by a field emission process. In
order to produce the high fields required for ion emission using
moderate extraction voltages (< 10kV), sharp, pointed emitters wetted
by a liquid metal are generally employed. Molecular SIMS studies
were made using a liquid metal ion source (19).
    Radioactive fission products travelling with very high energies
(MeV) have also been used to desorb molecular ions from a surface
(20).

Effects of Primary Beam Species Prediction of FABMS source sputtering yields can be generally supported by available data on secondary
ion emission coefficients. For secondary ions detected at zero angle
with respect to the surface normal, the secondary ion emission yield
generally increases with the mass of the primary ions because of the

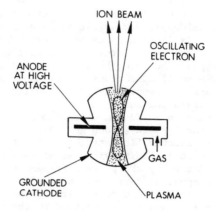

Figure 1.   Saddle Field Ion/Atom Gun.

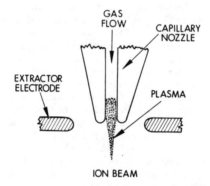

Figure 2.   Capillaritron Ion/atom Gun

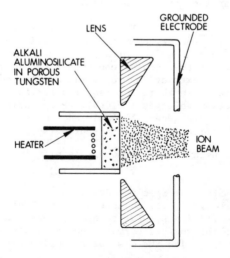

Figure 3.   Cesium Ion Gun

increased momentum transfer during the ion-target collisions. Argon
(40 amu) is often used as the primary beam gas because of its avail-
ability and low cost. Increased secondary ion yields were observed
under xenon (131 amu) atom/ion bombardment (21) which led to other
approaches such as non-gas-fed sources using cesium ions (133 amu)
with a mass closely matching that of xenon. Further efforts to increase the mass of the bombarding species has seen the use of a mercury (200 amu) ion beam obtained using a saddle-field ion source
(22). Mercury represents one of the high mass atomic ions generated
by conventional ion sources and was reported to produce an enhancement of the secondary ion signal. An attempt to exceed the mass
limitation imposed by monatomic elements has resulted in FABMS
studies involving molecular ions (23). Siloxane (diffusion pump
fluid) used with a saddle-field ion source, similar to the mercury
investigations, produced 531 amu ions whose bombardment gave a reported enhancement in the secondary ion yield.

Recent developments with the LMI source show promising results
for increasing secondary ion abundances (19). The primary ion
species investigated were of the type M_n^{z+} where M = Ga , In , SiAu
or Bi . Mass analysis of the beam composition of LMI sources reveal
the presence of cluster and multiply-charged ions, in addition to
the predominant M^+ ion. These complex multiply-charged species provide a means for increasing the bombardment energy without the need
for high accelerating voltages and thereby increasing the secondary
ion yield. Future approaches may see FABMS guns which bombard
organic sample surfaces by charged glycerol clusters in the range of
10Å ($\sim 10^5$amu) to 100Å droplets ($\sim 10^8$amu) .

Although it is clear from the above discussion that increased
mass of the primary ion species leads to high yields, nearly any ion
could be used. Successful FABMS spectra were obtained using nitrogen gas when inert gases were not available for the Capillaritron.
Even when gas from a tank is not available, which has occured at
trade show exhibitions, air from the room was used to obtain good
spectra as seen in Figure (4).

Primary Ion Flux Desorption rates increase with higher primary ion
flux rates until saturation effects set in. Saturation may be due
to depletion of sample, loss of matrix or replenishment not keeping
up to the desorption rate.

Assume the primary ion current leaving the gun is in the range
of one microampere. After travelling a distance of 10cm the current
density (j) is about $1\mu A/cm^2$ when the spread is about 5°. If
the target area (a) is $5mm^2$ then the primary ion beam (i) bombarding the target is given by

$$i = ja = \frac{(10^{-6}A)(0.05cm^2)}{cm^2} = 5 \times 10^{-8}A \qquad (1)$$

The number of primary ions per second (n) bombarding the target is

$$n = \frac{i}{e} = \frac{5 \times 10^{-8}A}{1.6 \times 10^{-19}coul} = 3.125 \times 10^{11}\frac{part}{sec} \qquad (2)$$

where $e = 1.6 \times 10^{-19}$ coul the ionic (or electronic) charge.
Typically the yield of secondary ions (Y), defined as the ratio of
secondary to primary ion current is about 1 percent. The secondary
ion current (I) is then

$$I = iY = 5 \times 10^{-8} \times 0.01 = 5 \times 10^{-10} A \qquad (3)$$

or 3.125×10^9 ions/sec .

Generally, the secondary ion current desorbed from a surface, bombarded by the primary ions, is composed of various masses that can be resolved by the analyzer. We will instead assume the sample is a
single mass of 100 amu. Consider a 100 nanogram of sample mass (m)
dissolved in a glycerol matrix of $\frac{1}{2}\mu\ell$ and placed on the target. The
total number of sample particles (N) is given by

$$N = \frac{m \, N_o}{M} = \frac{10^{-7} \; 6.023 \times 10^{23}}{100} = 6 \times 10^{14} \text{ particles} \qquad (4)$$

where m is the sample mass, N_o is Avagadro's number 6.023×10^{23}
particles per mole and M is the molecular weight of the sample in
grams/mole, or 100 amu in this example.

If the sample is uniformly dispersed in the glycerol with a density of 1.26, then there would be one sample molecule for every 6900
glycerol molecules. The FABMS process demonstrates a much higher
ratio of sputtered sample ions to glycerol ions, otherwise the signal
would be totally lost in the glycerol signal (noise). This leads to
the belief that a sample amenable to FABMS analysis probably resides
mostly on the glycerol surface. This is also seen in data showing
sample depletion with time accompanied by an increase in the glycerol
peaks.

Geometry and Mass Spectrometer Limitations

Under consideration is the geometry of the target surface in relation
to the incident primary beam and the direction to the analyzer. This
will be followed by an examination of limitations found on mass spectrometers along with solutions and suggestions to aid in converting
to FABMS.

Geometrical Arrangements   To arrange appropriate and/or optimal conditions for secondary ion emmission yield, the geometrical angles
will first be defined.

Figure 5 shows the incident primary beam impinging upon the target containing a liquid or solid sample. Angle of primary ion incidence upon the target surface is conventionally defined with respect
to the surface normal. The figure illustrates the typical FABMS
arrangement. The ion gun producing the primary beam is often at an
angle of 90° with the axis of the apertures to the analyzer. The
sample at the tip of the probe is introduced into the ion source
using the vacuum lock inlet. When the target is rotatable about the
axis perpendicular to the figure, the angle of incidence (θ) can be
varied. This also varies the angle to the analyzer (φ) since these
two angles are complementary (θ + φ = 90°) . It has been found that

the spectral maximum has an optimum value when θ = 64° (21, 25, 26).
For some spectrometers it is not possible to rotate the target about
that axis because of the position of the vacuum inlet, but it is
possible to attain this optimal angular arrangement. Figure 6 is a
three dimensional diagram showing the primary beam and the analyzer
at 90° from each other and three different target tip configurations
appropriate for different inlets. Tip shapes on each of the targets
that achieve the optimum angle discussed above are depicted.
Figure 7 shows a cross-sectional view of the target surface and how
each probe type fits the arrangement initially illustrated in
Figure 6. Each target has the same surface in common but enters
along a different axis. If the vacuum inlet is not along a major
axis, the target angle can still be made to fit the optimal angle.
Similar configurations can be analyzed even if the primary beam and
the analyzer do not make an angle of 90°. Of all the possible difficulties encountered in converting to FABMS capability, it is the
vacuum probe inlet that makes conversion possible and is the only
absolute requirement.

Pressure Problems Many of the older and some of the new smaller mass
spectrometers have relatively low pumping capacity. Often there is
no differential pumping to minimize the source load on the analyzer
chamber. Extra pumps have been successfully added to spectrometers
but this requires a major alteration to the spectrometer manifold.
The gas guns described previously produce a gas load such that the
operational pressure can exceed 10^{-4}torr. This will degrade the
spectra by decreasing the signal, especially for the higher mass
peaks. Figure 8 shows the peak size of two PFTBA mass peaks as a
function of pressure. The excess pressure was due to xenon gas bled
into the entire chamber of a benchtop mass spectrometer, including
the source and the analyzer. The pressure data was not corrected for
xenon but was made with an ionization gauge calibrated for nitrogen.
To correct the pressure these values should be divided by a factor
of 2.5. It would be desirable to operate below 10^{-4}torr and preferably 6 x 10^{-5}torr or lower as seen on the curves. Never the less
FABMS spectra can be taken at 2 x 10^{-4}torr, as shown in Figure 9
(27).
    FABMS can be accomplished on spectrometers having low pumping
speeds, using the Cesium Ion Gun. This gun causes no gas load and
produces virtually no neutral atoms as discussed previously. The
gun should be kept at a reasonable distance from the ionization
source to avoid heating which can limit the sample lifetime.

Flange Port Difficulties Several mass spectrometer types in use
have non-standard flanges for mounting the FABMS gun. Some do not
have conflat configurations and other flange ports are just too
small. Both difficulties can be solved using a flange adapter
having a conventional mating flange at one end to fit the gun and
the match to the spectrometer flange at the other end. Generally the
gun fits inside this adapter. In some cases this puts the gun far
from the source resulting in low beam current densities at the target. It is less of a problem with the Cesium Gun because it is more
amenable to focusing, which is not readily available to the gas
guns. A second solution is the use of the MiniFACS gun. Designed
to fit small flange openings, this gun configuration brings the ion

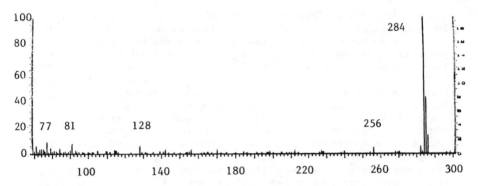

Figure 4. Spectrum of cetats [Hexadecyl trinethyl asmmonium
para - toluene sulphate $CH_3(CH_2)_{15}N^+(CH_3)_3$]. Taken during trade
show using air for the primary beam.

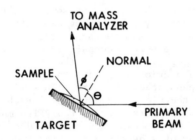

Figure 5. Diagram of the Incident Beam and Secondary Ions in
relation with the Target

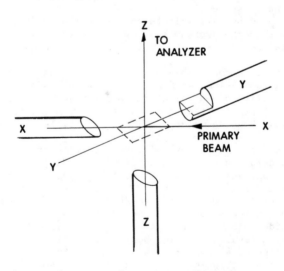

Figure 6. Several target configurations

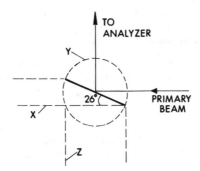

Figure 7.   Optimum angular relationship at the FABMS target.

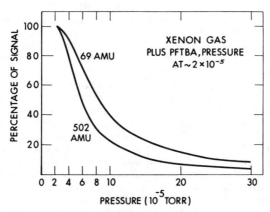

Figure 8.   Decrease of signal with increasing pressure on an HP 5995 Bench-Top Mass Spectrometer.

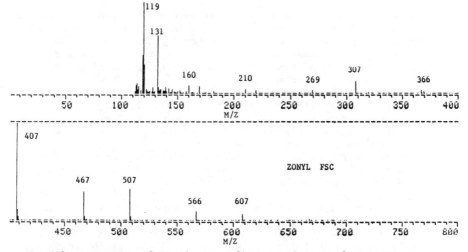

Figure 9.   Spectrum of Zonyl FSC a fluorosurfactant $[R_f CH_2 CH_2 SCH_2-CH_2 N^+ (CH_3)_3 CH_3 S^- O_4$ with $R_f = F(CF_2 CF_2)_{3-8}$. Taken on an HP 5995 Bench-Top M.S.

beam generator to within 1cm of the aperture on the mass spectrometer
ionization source. Thus high current densities are maintained even
when operating at low current levels.

No Flange Port Available   Some spectrometers have no flange port
available for the FAB gun. Never the less a FABMS conversion can be
made using the DIP Gun only if a vacuum inlet exists. The vacuum inlet must be larger than one-quarter inch in size, preferably one half-
inch. DIP Guns have been made and tested having diameters of 11mm,
0.478 inches and ½ inch. Of course the DIP Gun can be utilized and
is even desirable if a spare vacuum port is available because it is
simple to install and assures alignment between the gun and the
target (28).

The DIP Gun is an insertion probe which contains both the ion
gun and the target tip. This design, placing the ion generator at
the end of a narrow probe, was made possible using the Capillaritron
which has very small dimensions. Another benefit derived from the
Capillaritron in the DIP Gun is that FABMS operation can be performed
right after insertion and the gun can be removed immediately after
because neither heating nor cooling is required.

The details of the configuration at the end of the DIP Gun designed for a quadrupole is illustrated in Figure 10. The nozzle tip,
where the ion beam is generated, is located within and at the end of
the probe shaft. The ion beam passes axially through the probe tip
and impinges upon the target containing the sample. The target which
is electrically grounded, has been designed for quadrupole mass spectrometers. An electrically insulated target, shown in Figure 11, is
used with a magnetic sector mass spectrometer where ionization
sources are operated at several kilovolts. In this latter design,
the target holder contacts the ionization source block so as to
attain the same voltage level. Because the Capillaritron generates
fast ions and atoms, both equally capable of desorbing ions from the
sample surface, fast atoms will impinge upon the target even when the
gun voltage is lower than the ionization source voltage.

Special Gun Mounts   The guns described above are mounted on flanges or
or introduced through the vacuum lock. A few specialists have mounted the FABMS gun on the mass spectrometer ionization source structure
with good results. These mounting procedures require careful plan-
ning and alterations of the source structure to assure alignment and
minimal operational interference.

Rudat (29) has taken the nozzle from a Capillaritron gun and
mounted it on his spectrometer source. Because of the small distance
between gun and target, sufficient ion bombardment is attained at very
low gun currents. Closely related to this work is that of McEwen who
mounted a cesium gun on his ionization source (30). The gun function-
ed by using a filament to heat cesium chloride salts that drove off
the cesium. Before emerging from the gun toward the target, cesium
vapor is surface ionized by the filament, then accelerated by the
impressed moltage.

A different type of cesium gun was mounted on the ionization
source and compared with a conventional gas gun (10). This cesium
gun operates on the principles described earlier. The more favorable
results reported with the cesium gun test could be attributed in part
to the better focusing performed with the cesium gun and the shorter

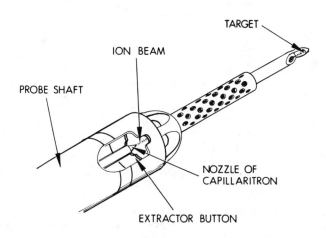

Figure 10.   Dip Gun tip for a quadrupole mass spectrometer.

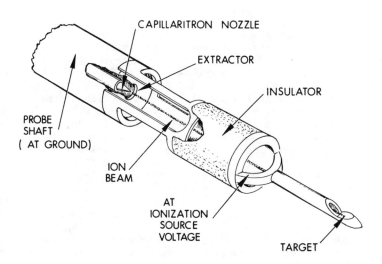

Figure 11.   Dip Gun tip for a magnetic sector mass spectrometer.

distance to the target, in addition to other features of the cesium
beam, discussed later.

Spectrometer Tests

Matrix Considerations   Non-volatile and thermally fragile molecular
samples are dissolved in a liquid matrix for introduction into the
spectrometer to perform FABMS measurements. The properties generally
imposed upon the matrix include: ability to dissolve samples and
possession of low vapor pressure to extend life time in the vacuum
environment. Sample life time should be several minutes to be able
to optimize the spectral signal and make several spectral runs. Many
organic liquids that satisfy these requirements have been used to obtain FAB data by several investigators (7, 31, 32). Two of the more
often used matrices are glycerol and thioglycerol   In addition, a
liquid metal matrix was used to float the sample to obtain FABMS
spectra (33).
   Glycerol, the most popular FABMS matrix, will be examined here.
Consider a sample dissolved in glycerol and a droplet placed on the
target. Assume a droplet volume of 0.5mm^3 with a density of
1.26g/cm^3 giving a mass of 6.3 x 10^{-4}g. The glycerol will evaporate
in the vacuum at a rate which is strongly dependent upon the temperature. The evaporation rate, ṁ , in g/sec, is given by

$$\dot{m} \, (\frac{g}{sec}) \; = \; 5.833 \times 10^{-2} P (\frac{M}{T})^{\frac{1}{2}} a$$

where P is the vapor pressure in torr, M is the molecular weight,
T is the Kelvin temperature and "a" is the area in cm^2. At 20°C the
vapor pressure is 1.75 x 10^{-4}torr and the evaporation rate is
2.9 x 10^{-7}g/sec from a sample whose area is 5mm^2. At this rate the
glycerol would last 14 minutes and at 40°C only 5 minutes. At 100°C
it would disappear in about 2 seconds. This illustrates the importance of maintaining a cooled target and source below 30°C. Heating
transfer tubes and analyzers should be avoided. Data can be taken at
higher temperatures, but the short lifetimes require excellent preparation and impose insufficient time for optimization.
   The next question is: what is the effect of sputtering on
sample lifetime? If we assume that the sputtering yield is unity and
the current is 5 x 10^{-8}A , the sputtering rate would be 3 x 10^{11}part/
sec or 4.6 x 10^{-11}g/sec. This is over 6000 times lower than the evaporation rate at 20°C. Even if the yield were higher and the ion
current raised, it is not likely that the sputtering rate would com-
pete with evaporation. However, the ion bombardment can heat the
sample and target structure, raise the temperature, and cause more
rapid evaporation.
   Ion bombardment of glycerol with sample dissolved can produce
molecular ions and glycerol cluster ions. Field (34) studied the
effects of bombarding a glycerol sample to study the products. Better
molecular ion spectra were found to occur when acid or alkali salts
were added (2).

Spectrometer Preparation   To operate with sufficient sample lifetime,
the ionization source and nearby components such as the transfer tube
and exit slits should be cooled hours before initiating tests. This

is to assure that the source temperature is below 30° to produce
sample lifetimes of over 10 minutes. The electron emitting filament
used with EI must be turned off. In some cases it helps to optimize
the spectrometer parameters to favor high mass sensitivity. Then it
should be balanced over the mass range after appropriate high mass
peaks are observed.

   To assure that proper gun operation and alignment exist, the
following testing procedure is suggested.
1. Turn on the FABMS Gun with no target in the source. The pressure
   in the source chamber will increase when a gas gun is used but
   show no effect with a cesium gun. The spectrum of the gas, be it
   xenon or argon, will be observed - or cesium from a cesium gun.
2. Insert the bare target into the source and turn on the ion gun.
   Spectra of the target material will be observed. This is
   actually a SIMS spectra. If the target material is stainless
   steel, the spectrum will show peaks in the 50 to 60 amu range for
   isotopes of iron and nickel. Optimize the position and angle of
   the target tip. Note that with the tip in place, the repeller
   voltage is generally ineffectual because of shielding. However,
   applying voltage to the target, if it is electrically isolated,
   can enhance the signal (26).
3. Place a droplet of glycerol on the target and insert the probe to
   observe the spectrum of the glycerol monomer, dimer, trimer, etc.
   and derivatives. Optimize the parameters to obtain strong peaks
   and minimum noise.
4. Dissolve a known sample in glycerol, place it on the target, insert and take the FAB spectrum. Compare the spectrum with published spectra and optimize the signals.
5. Then one is ready to measure an unknown sample with the knowledge
   that the FAB mode is in good operation.
This series of tests assures that: the gun is operating, the target
is in place, glycerol is in the correct place, a known FAB spectrum
is duplicated.

Noise   When the spectral signals are low, FABMS spectra show a con-
tinuous background noise. A peak will show up at virtually each
mass (32). This is thought to be caused by the very process of ion/
atom bombardment. When the energetic ions/atoms strike the sample
surface, momentum is transferred which causes the desorption of sputtered molecules and ions at perhaps a short distance away from the
impact site. But right at the impact site, the collision causes
fragmentation. Some of these fragments are charged and produce peaks
all over the spectral range. In addition, fast ion/atoms are reflected off the various surfaces and into the analyzer region. A magnetic sector spectrometer will not allow the atoms to reach the detector because of the curved path and the ions do not have the correct energy to pass through the curved path to produce a signal.
However, a quadrupole will transmit fast atoms and fast ions with no
discrimination which produces noise. One would expect that xenon,
more so than argon, would produce a lower noise level, or higher
signal-to-noise ratio because fewer primary ions/atoms are required
to produce a comparable signal. Cesium ions were found to be even
more favorable in producing low noise on a quadrupole, as discussed
previously and shown in Figure 12 (18). Much of the noise is attributed to fast atoms which are thought to undergo primary scattering
and more readily pass through the analyzer.

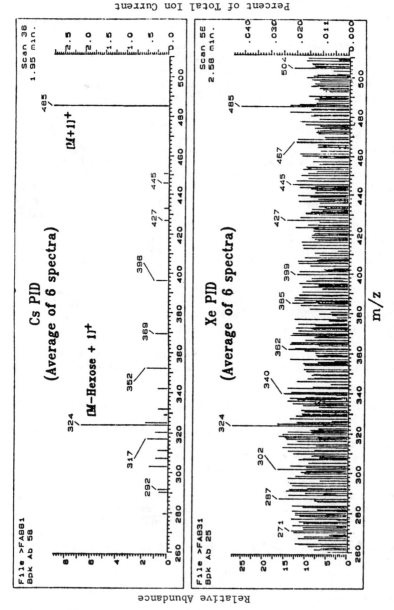

Figure 12. Spectrum of Kanamycin taken on an HP 5985 comparing a cesium with a xenon primary

Some New and Old Techniques

In considering the many possible arrangements of gun, target and
analyzer, the direction most avoided is "in-line". When the primary
beam is fired in line with the analyzer, the beam ions contribute
greatly to the noise background without even being scattered. However, two successful examples of near-in-line mounting are described
here.

DIP Gun In-Line on a Quadrupole A standard DIP Gun, described above,
was fitted with a special tip for testing on a Finnigan 4500 Mass
Spectrometer which has the vacuum lock inlet in line with the quadrupole analyzer. Requirements are imposed that the primary beam not
impinge directly on the analyzer entrance apertures and that the
secondary ions leaving the target be directed toward the analyzer.
Figure 13 shows a configuration that satisfies these requirements.
The ion beam leaving the Capillaritron nozzle is deflected downward,
by the proximity of the Extractor Tab, into the tip Cylinder. The
Target at the end of the tip contains the sample. Bombardment of the
sample at the grazing angle of incidence enhances the production of
slow molecular ions and also produces fast scattered ions/atoms. A
positive potential on the Shield/Repeller provides the field to repel
slow ions toward the analyzer while they are attracted toward the
Draw Out electrode. The Shield/Repeller also is the barrier preven-
ting primary ions/atoms from traveling directly toward the analyzer.
This general configuration without the voltage on the Shield/Repeller
and the deflector tab was tested using a DIP Gun on a 4500 and produced suitable spectra shown in Figure 14 (35, 36).

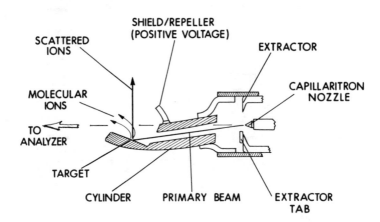

Figure 13.   DIP Gun tip for an inline vacuum lock inlet.

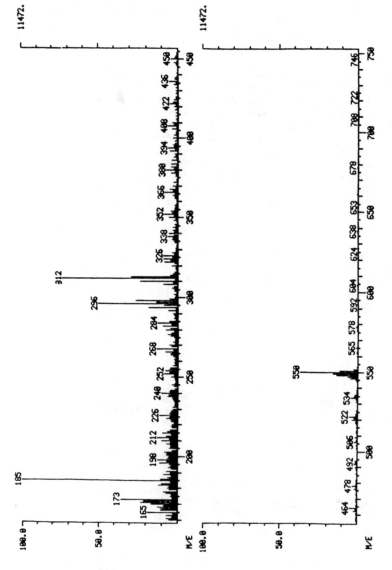

Figure 14. Spectrum of Dimethyl octadecyl ammonuum chloride
[(CH₃(CH₂)₁₇)₂N⁺(CH₃)₂]. Taken on a Finnigan 4500 using a Dip
Gun with the inline configuration.

Cesium Gun In-Line on a Magnetic Sector   A cesium gun was mounted in
the source manifold oriented so that the ion beam just misses the
entrance aperture of an MS-50 (37).   The target, coated with sample,
is the inside of a truncated cone.   The grazing angle of incidence
was also advantageous in producing more secondary ions.

Lowering Noise Levels   Techniques to lower the noise level due to
ion/atom scattering are being investigated.   The DIP Gun target is
oriented to face the analyzer, described before, which is the configuration found to maximize the secondary ion signals.   But it also
results in intense ions/atoms scattering toward the analyzer.   The
DIP Gun can be rotated to decrease the noise due to scattering.
Figure 15 shows the target tip facing the analyzer (0°) and rotated
through 90° and 180°.   At 90° few ions/atoms are directly scattered
toward the analyzer.   At 180° no ions/atoms are scattered toward the
analyzer.   The noise level decreases markedly as the angle increases
with a minimum at 180°.   The signal level was found, under some conditions, not to vary appreciably since the slow moving secondary ions
can be accelerated toward the analyzer by the repeller and draw out
electrode.   The signal-to-noise ratio was found to increase by a
factor of 40 on a quadrupole mass spectrometer (35).
    The second example employs the near grazing primary ion inci-
dent on the target which faces the analyzer entrance aperture of a
magnetic sector.   The scattered ions leave nearly parallel to the
surface and the slow molecular ions are drawn out using an immersion
lens (37).

Secondary Ion Mass Spectrometry   SIMS preceded and laid much of the
groundwork used in FABMS as discussed in the Background.   To come
full circle a FAB apparatus on a mass spectrometer can be used to

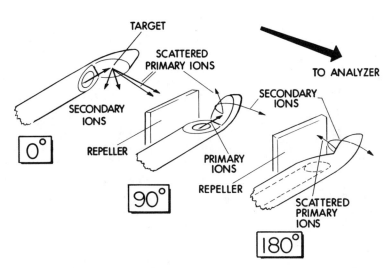

Figure 15. Dip Gun tip shown in three orientations of 0°,90°.
and 180° used to minimize FAB noise.

analyze surfaces of a wide range of materials , including the
target or in the form of small samples.   Both fast atoms and ions
are capable of desorbing ions from virtually any solid surface.
Similar to Organic SIMS, inorganic surfaces are analyzable. The target surface can be the sample as in the tests described previously
to determine the sensitive position and orientation of the target by
observing the SIMS spectra of the stainless steel target. Other
materials in the form of foils, powders or fabrics can be cemented
to the target and bombarded by the beam from the FABMS gun. Several
metals and ceramics were analyzed by this method.

A tantalum foil was cut into a sample slightly smaller than the
target stage. It was glued on with several different ordinary
cements. The resultant spectrum revealed a wide range of tantalum
and oxide molecules (clusters). These include three striking peaks
of Ta, TaO and TaO_2 followed by a series of much smaller peaks, each
separated by 16 amu. Also observed was Ta_2, Ta_2O, Ta_2O_2 and Ta_3,
Ta_3O, Ta_3O_2 etc. These were clear single peaks because there is only
one dominant isotope of tantalum. A molybdenum foil was also analyzed and this is shown in Figure 16. Molybdenum has seven dominant
isotopes between 92 and 100 and the relative abundances seen in the
spectrum is approximately equivalent to the isotope abundances.
Another grouping with oxygen added is seen 16 amu above the first one,
and a third with O_2 added is also observed. The Mo_2 group with the
0 and O_2 additions are also observable. Only slight evidence of the
Mo_3 group is seen.

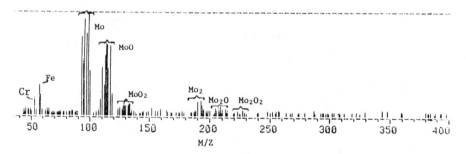

Figure 16. SIMS spectrum of a molybdenum foil taken on an HP 5995
Bench-Top MS.

Acknowledgment

The author wishes to acknowledge the help provided by John F. Mahoney in preparing this paper.

Literature Cited

1. Honig, R.E., "The Development of Secondary Ion Mass Spectrometry (SIMS)", 32nd Annual Conference on Mass Spectrometry and Allied Topics", Retrospective Lectures, San Antonio, Texas, 19, (27 May-1 June 1984).
2. McNeal, C.J., "Symposium on Fast Atom and Ion Induced Mass Spectrometry of Non-Volatile Organic Solids", Anal. Chem. 54, 43A, 1982.
3. Scheifers, S.M.; Hollar, R.C.; Busch, K.L. and Cooks, R.G., "Molecular Secondary Ion Mass Spectrometry", Amer. Laboratory, 19, March 1982.
4. Day, R.J.; Unger, S.E. and Cooks, R.G., Anal. Chem. 52, 557A 1980.
5. Benninghoven, A., Int. J. Mass Spectrom. Ion Phys., 46, p. 459, 1983.
6. Busch, K.L. and Cooks, R.G., Science, 218, p. 247, 15 October 1982.
7. Barber, M.; Bordoli, R.S.; Sedgwick, R.D.; and Tyler, A.N.; J.C.S. Dhem. Comm., p. 325 (1981).
8. Barber, M.; Bordoli, G.J.; Elliott, R.D.; Sedgwick, R.D. and Tyler, A.N., Anal. Chem., 54, p. 645A, 1982.
9. Barber, M., R.S. Bordoli, R.D. Sedgwick and A.N. Tyler, "Fast Atom Bombardment Mass Spectrometry, Anal. Chem. Symp. Ser., 12 pp. 177-83, 1983.
10. Aberth, W.; Straub, K.M.; and Burlingame, A.L., Anal. Chem., 54, pp. 2029-2093, 1982.
11. Campana, J.E., Int. J. Mass Spectrom, Ion Phys., 51, pp. 133-4, 1983.
12. Mahoney, J.F.; Perel, J.; S. Taylor, "Primary Ion Sources for Fast Atom Bombardment Mass Spectrometry", Am. Laboratory, p. 92, March 1984.
13. Franks, J., Inter. J. Mass Spectrom. Ion Phys., pp.343-6, 1983.
14. Mahoney, J.F.; Perel, J. and Forrester, A.T., Appl. Phys. Lett. 38(s), 320, 1981.
15 Mahoney, J.F.; Goebel, D.M.; Perel, J. and Forrester, A.T., Biom. Mass Spectr., 10, p. 61, 1983.
16. Perel, J. and Mahoney, J.F., "Analysis of the Operation of the FACS (Fast Atom Capillaritron Source)", Paper MOD4, ASMS 13th Annual Conference on Mass Spectrometry and Allied Topics, Honolulu, HI, 6-11 June 1982.
17. Aberth, W. and Burlingame, A.L., "Use of a Cesium Primary Beam for Liquid SIMS Analysis of Bio-organic Compounds", Springer Ser., Chem. Phys., 25, pp. 167-71, 1983.
18. Kenyon, C.N.; Goodley, J.; Mahoney, J. and Perel, J., "Particle Induced Mass Spectrometry Using a Conventional HP 598X Quadrupole MS with a Cesium Primary Ion Beam", 31st Annual Conference on Mass Spectrometry and Allied Topics, Boston, MA, p. 864, 8-13 May 1983.

19. Barofsky, D.F.; Giessmann, V.; Bell A.E.; and Swanson, L.W.,
     Anal. Chem., 55, p. 1318, (1983).
20. MacFarlane, R.D.; Acc. Chem. Res., 15, pp. 268-275, 1982.
21. Martin, S.A.; Costello, C.E.; and Biemann, Anal. Chem., 54
     pp. 2362-68, 1982.
22. Stoll, R; Schade, V.; Roellgen, F.W.; Giessmann, V. and Barofsky,
     D.F., Inter. J. Mass Spectrom. and Ion Phys., 43, pp. 227-229,
     1982.
23. Wong, D.S., Stoll, R. and Roellgen, F.W., Z. Naturforsch., 37A,
     pp. 718-719, 1982.
24. Mahoney, J.F.; Perel, J.; Goodley, P.C.; Kenyon, C.N. and
     Faull, K., "Modification of an HP 5985 GC/MS for FAB using a
     Fast Atom Capillaritron Source (FACS)", Inter. J. Mass Spectrom.
     and Ion Phys., 48, pp. 419-422, 1983. Presented at 9th Inter.
     Mass Spectrom. Conf., Vienna, August 1982.
25. Faull, K.F.; Tyler, A.N.; Sims, H.; Barchas, J.D.; Massey, I.J.;
     Kenyon, C.N.; Goodley, P.C.; Mahoney, J.F. and Perel, J, Anal.
     Chem., 56, p. 308, 1984.
26. Caprioli, R.M.; Bechner, C.F., and Smith, L.A., Biomed. Mass
     Spectrom., 10 pp. 94-97, (1983.
27. Perel, J.; Mahoney, J.F., and J-L. Truche, "FAB for Bench-
     Top and Orphaned Mass Spectrometers", Paper No. 409, Pittsburgh
     Conference and Exposition, New Orleans, L.A., 25 Feb-1 March
     1985.
28. Perel, J.; Faull, K.F.; Mahoney, J.F.; Tyler, A.N. and
     Barchas, J.D., "Direct Insertion Probe Gun for FAB Mass Spectro-
     metry", American Laboratory, November 1984.
29. Rudat, M.A.; Anal. Chem., 54, p. 1917, 1982.
30. McEwen, C.N.; Anal. Chem., 55, pp. 967-8, 1983.
31. Heller, D.N.; Fenselau, C.; Yergey, J., and Cotter, R.J., Anal.
     Chem., 56, pp. 2274-2277, 1984.
32. Rinehart Jr, K.L., Science, 218, p. 254, 15 October 1982.
33. Ross, M.M., and Colton, Richard J., Anal. Chem., 55, pp. 1170-1,
     1983.
34. Field, F.H., J. Phys. Chem., 86, pp. 5115-23, 1982.
35. Perel, J.; Mahoney, J.F., "Lowering FAB Noise in Quadrupole Mass
     Spectrometers", 33rd Annual Conference of Mass Spectrometry and
     Allied Topics, San Diego, CA, 26-31 May 1985.
36. Voyksner, R.; private communication.
37. Alberth, W., and Burlingame, A.L., Anal. Chem., 56, p. 2915,
     1984.

RECEIVED July 24, 1985

Fast Atom Bombardment Secondary Ion Mass Spectrometric Surface Analysis

J. A. Leys

Analytical and Properties Research Laboratory, Central Research Laboratories, 3M, St. Paul, MN 55144-1000

Fast atom bombardment (FAB) secondary ion mass spectrometry (SIMS) permits the static and dynamic surface analysis of insulating materials without the severe charging effects experienced by ion beam bombardment. The relatively low potential surface charge which may accumulate on insulating materials during fast atom bombardment due to secondary ion and electron emission is readily neutralized by low energy photoelectrons produced by irradiation of the area adjacent to the sample with a small mercury discharge lamp located within the analysis chamber. Characteristics of the FAB gun and applications of the FAB SIMS technique to the analysis of polymer films and inorganic powders are described.

SIMS analysis of electrically insulating surfaces using a positive or negative ion beam requires surface charge neutralization in order to obtain useful secondary ion yields. This is most often accomplished by irradiating the surface under investigation with a 0.5 to 3 keV electron beam. This requires very careful adjustment of neutralizing beam parameters since any small residual charge on the specimen surface may affect both the secondary ion yield and the energy distribution of the secondary ions. In quadrupole mass analyzer based systems, slight surface charging often results in the secondary ion energy distribution shifting to values outside the band pass of the quadrupole prefilter. Under these conditions, a partial or total loss of signal occurs. In addition, as the specimen is eroded by sputtering, the work function and other surface characteristics change resulting in the need for more or less constant fine tuning of the neutralization parameters during surface analysis. Further, the added thermal and electrical effects of the electron beam can cause damage to fragile polymeric materials, it can result in unwanted electron beam stimulated surface desorption, and

0097-6156/85/0291-0145$06.00/0
© 1985 American Chemical Society

can cause migration or diffusion of alkali metals away from the surface.

The use of FAB probes to alleviate the problems associated with surface charging has been previously reported in literature (1-4). In general, these articles suggest that no external neutralization source is required with a neutral beam on insulators. This article will show where neutralization can be used to advantage and will describe results obtained with a simple scheme of neutralization utilizing photoelectrons ejected from the area surrounding the sample when irradiated by a small mercury discharge lamp within the analysis chamber. The use of FAB SIMS for the analysis of polymer surfaces and inorganic oxides will also be reported.

Experimental

FAB sources of various types are described in literature. The high "current" sources such as the Saddle Field (5) or Capillitron (6) which have recently come into use for bulk mass spectrometric analysis of involatile organic materials are generally not useful for "low-damage" surface analysis. The reason is because of the inability of these sources to obtain steady or reproducible atom flux at the low dose conditions, i.e., 10^{-6} to 10^{-10} A/cm^2 equivalent, required for the analysis of thermally sensitive materials. The source used in this study is shown in Figure 1 and consists of a commercial ion gun (Mini Beam - Kratos, Inc.) to which has been added a charge exchange chamber to neutralize the ions from the ion gun. Sources of this type have recently been described in literature (7-9). Since significant surface chemical differences between samples can be accompanied by only subtle intensity changes in molecular SIMS fragmentation patterns, emphasis was placed on designing a source which produced a stable and highly reproducible atom beam. The design shown permits precise measurement and control of the gas pressure in the ion source region and in the exchange chamber which are necessary requirements for reproducible operation. Figure 2 shows that for argon, the most efficient charge exchange occurs within a narrow pressure region centered at about 1.5 x 10^{-2} torr. The data show that the pressure at which the neutral flux is maximum is independent of the beam energy. This is due to the essentially constant charge exchange cross section for argon within this energy range. At pressures higher than optimum for neutralization, the neutral flux is reduced by scattering processes. At the optimum pressure, our measurements show that approximately one half of the ions are neutralized. The pressure in the ion source region is maintained at 1 x 10^{-4} torr and in the charge exchange chamber at 1.5 x 10^{-2} torr by automatic pressure controllers (Veeco APC/1000) which operate piezoelectric leak valves. When operated as an ion source (charge exchange chamber evacuated), the gun is capable of maximum argon ion current of 1.5 microamps at 5 keV at a spot size of about 2 mm. The maximum equivalent neutral flux as measured indirectly by secondary ion yields from neutral bombardment of a titanium metal target is approximately one half of the ion current.

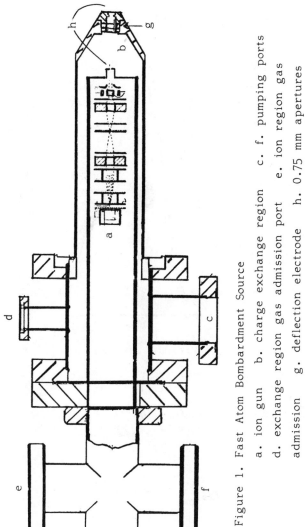

Figure 1. Fast Atom Bombardment Source

a. ion gun   b. charge exchange region   c. f. pumping ports
d. exchange region gas admission port   e. ion region gas
admission   g. deflection electrode   h. 0.75 mm apertures

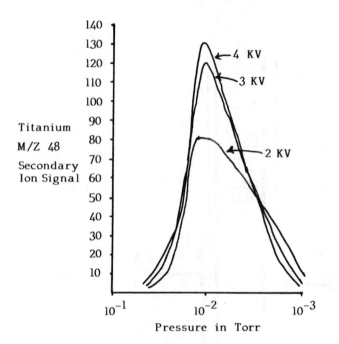

Figure 2. Argon Neutral Yield* vs. Charge
Exchange Chamber Pressure

*As measured by secondary ion signal at
M/Z 48 from titanium metal target

The secondary ion extraction optics and ion energy filter is of commercial design (Kratos). Also incorporated in the analysis chamber is a scanning electron gun having a 1 micron beam size which is used for imaging and positioning the specimen under the neutral beam. In most cases, the area where the FAB beam impinges upon the sample is visible by secondary electron imaging thus providing a convenient method of sample positioning. The charge neutralizer consists of a Series 81 Analamp mercury discharge lamp (BHK, Inc., Monrovia, CA), located within the analysis chamber approximately 7 cm from the specimen surface. The lamp is mounted on a 1 cm diameter copper bar which in turn is mounted on a chamber flange. This heat-sink mounting is required to prevent excessive pressure buildup in the lamp which causes an unstable discharge.

Results and Discussion

Charge Neutralization. When an insulator is bombarded with a beam of ions, the resultant surface charge will depend on the sum of the primary current and its potential, current from the emission of secondary ions and electrons, and tertiary currents representing miscellaneous contributions from scattered ions, neutrals, and electrons within the analysis chamber. When neutral particles are used for bombardment, the resultant surface charge will depend primarily on the sum of the secondary and tertiary currents. The surface potential buildup is thus much lower with a neutral beam than with an ion beam, and it is always possible to obtain a SIMS spectrum without any form of charge neutralization which is not the case when ion bombardment is used. Nevertheless, the charge effect with neutral bombardment can result in significant problems in SIMS analysis. Specifically, quadrupole mass spectrometers require that the secondary ion energy distribution be in a range generally below 20 eV in order to achieve efficient mass filtering. Although in principal one can decelerate secondary ions to below this potential; the difficulty with this approach is that the charge on the insulator is generally not maintained constant during analysis. This is particularly true of powdered materials, individual particles of which may achieve significantly different electrostatic charge. In addition, it is often advantageous to select a narrow energy band pass of secondary ions in order to differentiate between mass spectral interferences caused by molecular and atomic species (10). Under these conditions, changes or instabilities in the energy distribution of the secondary ions during the analysis will confuse the results.

Figure 3a shows a pattern of the FAB SIMS spectra of a photochromic glass. Molecular fragment ions are apparent from M/Z 60-70, and zirconium atomic fragment ions are observed at M/Z 90-96. No neutralization was used, and the band pass of the quadrupole prefilter was adjusted for maximum signal at 90 AMU. A direct measurement of the secondary ion energy distribution was not made, but computer modeling of the prefilter characteristics suggests that under the conditions used, the band pass was centered at about 45 eV. Even though the quadrupole resolution setting was at its optimal value, the resolution of the spectrum is poor because of the high energy of the secondary ions. Figure 3b shows the spectrum of the same glass after turning on the mercury discharge lamp and readjusting the prefilter settings to again obtain maximum signal at 90 AMU. (The center of the prefilter band pass was estimated to be 7 volts.) The y intensity axis in both cases was the same.

The neutralization effect of the mercury lamp appears to be due to the ejection of low energy photoelectrons from the vacuum chamber walls and equipment contained therein. Focusing an external source of mercury radiation directly onto the sample surface (using UV transmitting optics) does not result in charge neutralization. However, when this external radiation is directed to a small piece of stainless steel immediately above and to the side of the sample, then neutralization takes place. Figure 4 shows the target current from photoelectrons, obtained with a conductive target, as the target potential is varied from 0 to +20 volts. Over one half of the available photoelectrons are collected at a target bias of 5 volts. This suggests that an insulating specimen surface which may charge to +45 volts during fast atom bombardment would be a very efficient collector of the photoelectrons. Considering the secondary and tertiary currents normally involved during analysis, it is likely that the surface potential of insulators remains at 5 volts or below because of the "self-regulating" effect of charge neutralization by this technique. Charge neutralization can also be accomplished by irradiating the sample with low energy electrons from a source such as a heated tungsten filament. This technique adds considerable thermal energy to the sample unless the source is designed to shield the sample from the infrared radiation of the filament. High energy electrons such as from an electron gun having rastering capabilities can also be used, however, this requires careful and continual adjustment because neutralization is not self-compensating, and slight surface changes result in over or under neutralization.

Figure 5 shows the effect of thermal damage on polyethylene when utilizing a 2 keV, 1 microamp electron beam for charge neutralization. "Carbonization" of the surface begins to take place within the first few seconds of analysis when the electron beam is used whereas with the mercury-lamp-photoelectron method, a stable spectrum is obtained for periods of several minutes.

During analysis using negative secondary ions, the extraction optics and quadrupole detector are biased in such a manner as to pass and detect electrons as well as negative ions. Figure 6a shows the spectrum of contaminated quartz from argon ion bombardment with electron gun (2 keV) charge neutralization, and Figure 6b shows a spectrum of the same sample with equivalent argon neutral beam bombardment and without charge neutralization. The difference in background counts shown at 45 AMU is about a factor of 30.

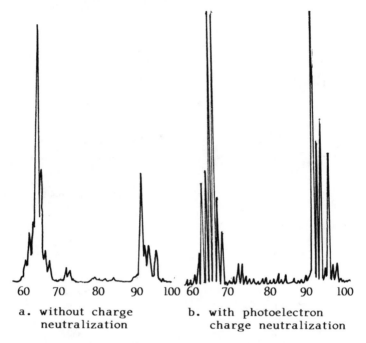

a. without charge
   neutralization

b. with photoelectron
   charge neutralization

Figure 3. FAB SIMS Spectra of Photochromic Glass

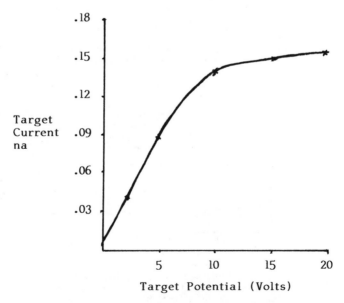

Figure 4. Target Current vs. Target Potential
          Mercury Discharge Lamp Induced
          Photoelectrons

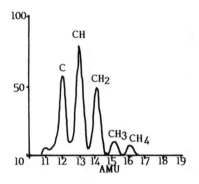

Figure 5a. First Spectrum, 10 sec
Scan, e Beam Neutralization

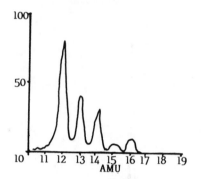

Figure 5b. 40th Spectrum, 400 sec,
e Beam Neutralization

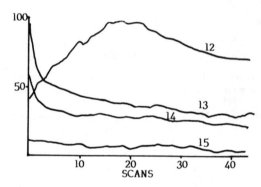

Figure 5c. Depth Profile of Polyethylene
Surface – e Beam Neutralization

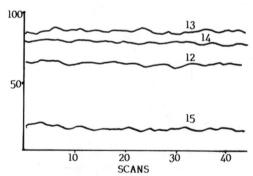

Figure 5d. Depth Profile of Polyethylene
Surface – Mercury Radiation
Induced Photoelectron
Neutralization

Figure 5. FAB SIMS Spectra of Polyethylene Showing Thermal
Damage with e Beam Charge Neutralization

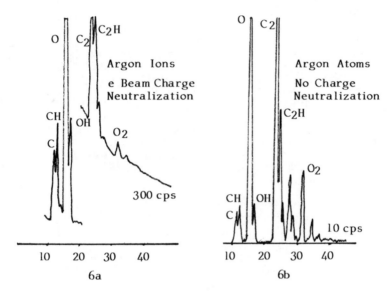

Figure 6. Negative SIMS Spectra of Contaminated Quartz

Cobalt Oxide Analysis. The magnetic characteristics of ferric oxide particles of the type used in magnetic recording media are often modified by surface treatments involving deposition of up to a few hundred angstroms of oxides of cobalt on the iron oxide particles. The cobalt oxidation state (oxide stoichiometry) is not readily determined on surfaces of this type by the more conventional techniques such as X-ray or electron diffraction or X-ray photoelectron spectroscopy (XPS or ESCA). These particles are thermally unstable and are electrical insulators. Initial attempts to differentiate the various oxides of cobalt with ion beam bombardment and electron beam charge neutralization were not successful. Table I shows the relative abundances of secondary ions from argon atom bombardment for the oxides and hydroxide of cobalt indicated. The data were normalized to the M/Z 59 intensity. Line 5 shows the average of the data shown in line 1 (CoO) and line 2 (Co$_2$O$_3$) and is what one might expect for an equal mixture of CoO and Co$_2$O$_3$ if the secondary ion yields were the same for both. This agrees well with the data for Co$_3$O$_4$ which can be thought of as CoO·Co$_2$O$_3$. The hydroxide spectrum is distinguished in particular by the large peak at M/Z 134. The negative ion spectrum of Co(OH)$_2$ also contains a parent peak at 77 which is not observed for the oxides. In addition, the hydroxide contains larger peaks at M/Z 33 (HO$_2$)$^-$ and 17 (OH)$^-$ than the cobalt oxides.

Table I. FAB SIMS Fragmentation of Cobalt Oxides

	Co$^+$ 59	CoO$^+$ 75	Co$_2^+$ 118	Co$_2$O$^+$ 134	Co$_2$O$_2^+$ 150
1. CoO	100	3.4	11.4	23	1.2
2. Co$_2$O$_3$	100	1.9	11.0	14	0.2
3. Co$_3$O$_4$	100	2.7	11.2	18	0.5
4. Co(OH)$_2$	100	7.6	11.7	95	4
5. CoO + Co$_2$O$_3$	100	2.6	11	18	0.7

Polymer Surface Analysis. The major technique used for the surface analysis of polymers has been X-ray photoelectron spectroscopy (XPS or ESCA). However, this technique is often not adequate to determine the molecular structure of polymers. This has prompted many workers to explore the potential of SIMS for this work (11-16). Significant problems encountered with ion beam bombardment in conjunction with electron beam charge neutralization have been drift in the polymer surface potential and thermal damage from the combined effects of the electron and ion beams. These problems do not exist when utilizing FAB in conjunction with photoelectron charge neutralization.

Figures 7a, b, and c show the SIMS spectra of known polymers PTFE, FEP, and PVF$_2$. Briggs and Wootton (13) report that radiation from electron beam charge neutralization

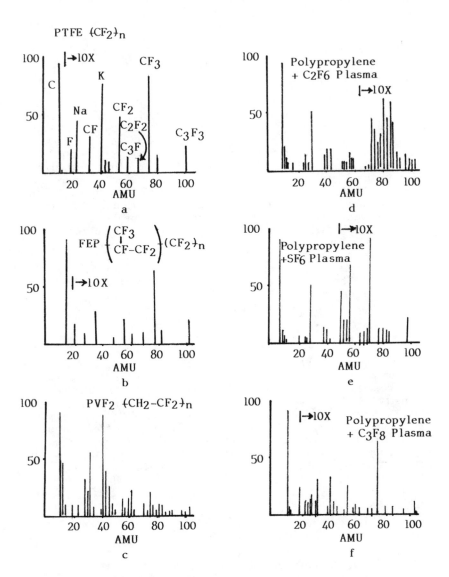

Figure 7. FAB SIMS Spectra of Fluoropolymers and Plasma Treated Polypropylene Surfaces

in itself caused significant emission of molecular ions such as CF_3^+ from surfaces such as PTFE. This was confirmed in this laboratory. No desorbed ions were observed with the mercury photoelectron method of charge neutralization. Figures 7d, e, and f show the SIMS spectra of polypropylene which has been surface modified by plasma treatment in the fluoride gas indicated. The peak intensity values from 12 through 35 AMU were normalized to carbon M/Z 12 intensity and are tabulated in Table II. SIMS spectra of PTFE and FEP do not show the hydrocarbon fragmentation series characteristic of the other surfaces because the hydrogen atoms are fully substituted by fluorine. Mass 13 for these samples shows 2% of the mass 12 peak. Carbon 13C accounts for 1% of this; the remaining 1% is likely due to surface contamination. The first difference between these two samples occurs at M/Z 69 (CF_3^+) where this peak is about twice the intensity for the FEP samples as the PTFE surface. At higher M/Z values, additional differences are observed. The relative intensities of peaks M/Z 13CH+, 14CH$_2^+$, and 15CH$_3^+$ for the plasma treated polypropylene surfaces are an indication of the amount of fluorine substitution. This is confirmed by the XPS or ESCA carbon (15) spectra of the same samples which are shown in Figures 8a, b, and c. The intensity of the XPS hydrocarbon peaks show general agreement with the intensity of the peaks at M/Z 13, 14, and 15 in the SIMS spectra.

The negative FAB SIMS spectra of the fluoropolymers did not contain any particularly useful information. The major peak observed is at M/Z 19 F⁻. Small peaks of one percent intensity or less are observed at M/Z 12 C⁻, 16 O⁻, 24 C$_2^-$, 31 CF⁻, and 38 F$_2^-$. Our experience suggests that negative FAB SIMS appears to be the most useful for the determination of the relative amount of oxygen in the polymer structure since positive SIMS has low sensitivity for oxygen and potential interference from hydrocarbons at M/Z 16.

Conclusion

FAB SIMS in conjunction with charge neutralization utilizing mercury discharge lamp induced photoelectrons permits low-damage highly reproducible analysis of electrically insulating surfaces. Stable spectra can be obtained from polymeric materials such as polyethylene for periods of an hour or more. Minor spectrum differences between samples such as the various oxides of cobalt, which may have previously been due to thermal or surface potential effects, can now be more confidently assigned to compositional differences.

Table II. Normalized Peak Intensities for the Spectra shown in Figure 7

AMU	PTFE $(CF_2)_n$	FEP CF_3 $(CF-CF_2)(CF_2)_n$	PVF$_2$ $(CF_2-CH_2)_n$	C$_2$F$_6$ Plasma on Polypropylene	SF$_6$ Plasma on Polypropylene	C$_3$F$_8$ Plasma on Polypropylene
12 C	100	100	100	100	100	100
13 C_{13}CH	2	2	64	28	14	9
14 CH$_2$			49	15	7	3
15 CH$_3$			11	7	3	1
16 CH$_4$, O						
17						
18						
19 F	1	1	10	4	3	2
20						
21						
22						
23 Na	4		11		6	1
24 C2			4		<1	<1
25 C$_2$H			8	2	1	<1
26 C$_2$H$_2$			20	6	4	1
27 C$_2$H$_3$			35	15	7	1
28 C$_2$H$_4$			14	2	1	<1
29 C$_2$H$_5$			24	8	3	<1
30 C$_2$H$_6$			8	8	7	
31 CF			60	58	46	35
32 CHF			5	2	1	
33 CH$_2$F			10	1		
34						
35						

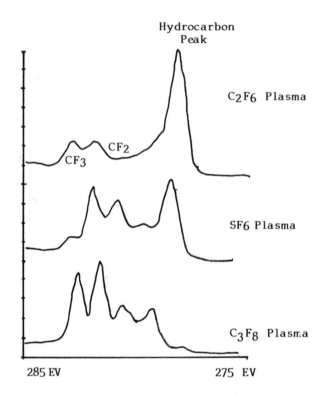

Figure 8. XPS Carbon (15) Spectra of
Plasma Treated Polypropylene

Literature Cited

1. Surman, D. J.; VanderBerg, J. A.; Vickerman, J. C. SIA Surf. Interface Analysis 1982, 4, 160-7.
2. Ilno, A.; Mizuike, A. Bull. Chem. Soc. Japan 1981, 54, 1975-7.
3. Wakefield, C. J.; Hazelby, D.; Taylor, L. C. E.; Evans, S. Int. J. Mass Spectrom. Ion Phys. 1983, 46, 491-4.
4. Borchardt, G.; Franek, H.; Scherrer, H.; Scherrer, S.; Weber, S. Int. J. Mass Spectrom. Ion Phys. 1983, 46, 507-10.
5. Franks, J. Int. J. Mass Spectrom Ion Phys. 1983, 46, 343-6.
6. Mahoney, J.; Perel, J.; Forrester, A. Appl. Phys. Lett. 1981, 38 (5), 320-2.
7. Surman, D. J.; Vickerman, J. C. Applications of Surface Science 1981, 9, 108-21.
8. Klaus, V. N. Vakuum Technik 1982, 31, 106-8.
9. Barber, M.; Bordoli, R.; Sedgwick, R.; Taylor, A. Nature 1981, 293, 270-5.
10. Blattner, R.; Evans, C. "Scanning Electron Microscopy IV"; SEM, Incorporated: Chicago, 1980.
11. Gardella, J. A.; Hercules, D. M. Anal. Chem. 1980, 52, 226.
12. Gardella, J. A.; Hercules, D. M. Anal. Chem. 1981, 53, 1879.
13. Briggs, D.; Wootton, A. B. Surf. Interface Anal. 1982, 4, 109.
14. Briggs, D. Surf. Interface Anal. 1982, 4, 151.
15. Briggs, D. "Ion Formation from Organic Solids II"; Benninghoven, A., Ed.; Springer Series in Chem. Phys., Springer-Verlanz, 1983; Vol. 25, p. 156.
16. Briggs, D. Surf. Interface Anal. 1983, 5, 113.

RECEIVED April 16, 1985

Secondary Ion Mass Spectrometry:
A Multidimensional Technique

Richard J. Colton[1], David A. Kidwell[1], George O. Ramseyer[2], and Mark M. Ross[1]

[1] Chemistry Division, Naval Research Laboratory, Washington, DC 20375-5000
[2] General Electric Company, Syracuse, NY 13221

This paper discusses SIMS as a multi-dimensional technique for the analysis of inorganic and organic materials. The paper is divided into two parts: inorganic SIMS and organic (molecular) SIMS. The inorganic SIMS part focuses on methods of quantitative analysis and depth profiling applications. In particular, the parameters that makes SIMS difficult to quantify -secondary ion yield, matrix effects, and instrumental effects - are reviewed as well as the various physical models and empirical methods used to quantify SIMS data. The instrumental and experimental parameters that affect SIMS depth profiling are also reviewed. The organic SIMS part discusses the method of ionization and the various sample preparation and matrix-assisted procedures used for analysis. The matrices include various solid-state and liquid matrices such as ammonium chloride, charcoal, glycerol, and gallium. A neutral beam source was developed to analyze thick, insulating films. Various chemical derivatization procedures have been developed to enhance the sensitivity of molecular SIMS and to selectively detect components in mixtures.

Bombarding a solid surface with low energy (keV) ions or neutrals results in the emission of secondary particles: positive and negative ions, neutrals, electrons, and photons. This phenomenon, known as sputtering, is dependent on several important parameters such as the energy, mass, and angle of the incident beam and the mass, structure, and binding energy of the atoms which form the surface of the target [1]. Mass analysis of the sputtered secondary ions forms the basis of secondary ion mass spectrometry (SIMS) [2].

0097-6156/85/0291-0160$09.50/0
© 1985 American Chemical Society

As a surface analytical tool, SIMS has several advantages over X-ray photoelectron spectroscopy (XPS) and Auger electron spectroscopy (AES). SIMS is sensitive to all elements and isotopes in the periodic table, whereas XPS and AES cannot detect H and He. SIMS also has a lower detection limit of $\sim 10^{-5}$ atomic percent (at.%) compared to 0.1 at.% and 1.0 at.% for AES and XPS, respectively. However, SIMS has several disadvantages. Its elemental sensitivity varies over five orders of magnitude and differs for a given element in different sample matrices, i.e., SIMS shows a strong matrix effect. This matrix effect makes SIMS measurements difficult to quantify. Recent progress, however, has been made especially in the development of quantitative models for the analysis of semiconductors [3-5].

SIMS methodology has evolved along two distinct lines. The first and original method showed SIMS as an analytical tool for depth profiling and microanalysis. Specialized instruments with microscope or microprobe capabilities were developed for depth profiling, ion imaging and micro-area analysis [2,6]. This SIMS method, commonly referred to as "dynamic SIMS", uses a relatively high primary ion beam flux ($> 1\times10^{-6}$ A/cm^2) to generate specimen sputtering rates of > 50 Å/min. The high sputtering rates enhance the sensitivity of the method. The dynamic SIMS method has been applied primarily to studies in electronic technology and material science [7-9].

The second SIMS method was pioneered by A. Benninghoven (Univ. of Münster, West Germany) in the late 1960's and is capable of analyzing surface monolayers [10]. To achieve monolayer sensitivity, it is first necessary to reduce the sample sputtering rate by lowering the primary ion beam flux (typically $< 1\times10^{-9}$ A/cm^2) and second, in order to compensate for the corresponding loss in signal intensity (due to the lower flux), the analysis area is increased by broadening or rastering the primary ion beam. This SIMS method known as "static" [12,13] or low damage [14] SIMS has been applied to the study of gas-surface interactions [7-9,15-17]. (The pioneering work of Macfarlane (Texas A&M University) dealing with Californium-252 plasma desorption mass spectrometry was important in establishing that large organic molecules could be desorbed as intact molecular and molecular-like ions [11]).

As a sub-element of the static SIMS methodology, SIMS has become (most recently) a new ionization source for the analysis of nonvolatile and thermally labile molecules including polymers and large biomolecules such as proteins. Since most of these latter studies deal with the emission of polyatomic or molecular ions, the name "molecular SIMS" has been applied [18-21].

The application of molecular SIMS as a sensitive ionization source for nonvolatile and thermally labile molecules compares favorably with other new ionization methods in mass spectrometry such as field desorption (FD), Californium-252 plasma desorption (PD), fast heavy ion induced desorption (FHIID), laser desorp-

tion (LD) or laser microprobe mass analysis (LAMMA), and
fast-atom bombardment (FAB) or liquid SIMS [21-23]. In each of
these techniques, the molecules are desorbed and ionized
directly from the solid state and appear as molecular and/or
molecular-like (protonated, deprotonated, and/or cationized)
ions, e.g., $M^{\pm}\cdot$, $[M \pm H]^{\mp}$ and $[M + cation]^{+}$.

This paper discusses SIMS as a multi-dimensional technique
for the analysis of inorganic and organic materials. The paper
is divided into two parts: inorganic and organic (or molecular)
SIMS. The inorganic SIMS part focuses on the methods of
quantitative analysis and depth profiling applications. In
particular, SIMS matrix effects are defined and the physical
models and empirical methods used to quantify SIMS results are
reviewed.

The emission of molecular ions in organic SIMS is discussed
with respect to the method of ionization and the various sample
preparation and matrix-assisted procedures used. The matrices
include various solid-state and liquid matrices such as ammonium
chloride, charcoal, glycerol, and gallium. A neutral beam
source is described to analyze thick insulating films. Various
chemical derivatization procedures have been developed to
enhance the sensitivity of molecular SIMS and to selectively
detect components in mixtures.

Inorganic SIMS

The results discussed in this section deal primarily with the
methods used to quantify dynamic SIMS results obtained from
depth profiling studies of inorganic materials such as semiconductors.

Quantitative Analysis. SIMS has many unique features (compared
to other surface analytical techniques) such as hydrogen and
isotope detection, a detection limit as low as 10^{-14}g, surface/
monolayer sensitivity, compound specificity, and high spatial
resolution (<1000Å). In addition, SIMS is perhaps the most
versatile surface analytical technique. It can depth profile
semiconductors, image biological tissues, study gas-surface
interactions/adsorbate geometry, and ionize nonvolatile and
thermally labile molecules. However, it is often said that
"SIMS is difficult to quantify". While this statement is true
to a large extent, recent progress has been made to develop
quantitative methods that convert measured secondary ion
intensities to atomic concentrations and to apply these methods
to analyze depth profiles and ion images. The quantitative
methods used in SIMS have been reviewed recently by Werner [3],
Wittmaack [4] and Morrison [5]. Magee and Honig [24], Hofmann
[25,26] and Wittmaack [27] have reviewed the application of
these methods to depth profile analysis.

An example of a SIMS depth profile is shown in Figure 1 for a ^{11}B implanted Si specimen. The profile is plotted after a fashion being promoted by the ASTM E-42.06 Surface Analysis subcommittee on SIMS. The depth profile is recorded by monitoring the secondary ion intensity (in counts per second) of the ^{11}B$^+$ implanted species and the ^{30}Si$^+$ matrix element as a function of sputtering time (in seconds). The plot of secondary ion intensity versus sputtering time represents the raw experimental data which are converted to the ^{11}B atomic concentration (in atoms/cm^3) and depth (in nanometers). The concentration scale is obtained using a standard reference specimen (as will be described below); and the depth scale is obtained by measuring the depth of the sputtered crater by profilometric or interferometric techniques. (Assigning a linear depth scale assumes a linear sputter rate which may not apply. See below for a discussion of nonlinear sputtering effects.)

Let us begin by looking at the parameters that make SIMS measurements difficult to quantify. First, the secondary ion yield (defined as the ratio of the number of secondary ions sputtered from the surface of a solid sample to the number of primary ions incident upon the specimen) varies over four orders of magnitude from element to element. Second, the yield of each ion is affected by the composition of the matrix. This is the well known SIMS "matrix effect". Third, instrumental effects and ion collection/detection efficiencies can vary from instrument to instrument and specimen to specimen.

Secondary ion yields are defined as the total number of secondary ions sputtered from the specimen per incident ion of given mass, energy, charge, and angle of incidence [27]. Positive secondary ion yields are enhanced by oxygen and other electronegative reactive species and negative secondary ion yields are enhanced by cesium and other electropositive reactive species [4]. Depending upon the sample matrix, reactive materials have increased the secondary ion yields of some elements by orders of magnitude [29]. The energy of the primary ions, the mass of the primary ions, and the sputtering yield determine the concentration of the primary ion that is implanted into the sample matrix once equilibrium between the rate of implantation and sputtering is reached [30]. The nature and concentration of the implanted species also affects directly the secondary ion yields.

Anderson and Hinthorne [31] have determined that for positive secondary ions an element's secondary ion yield is inversely proportional to the first ionization potential of an element. A similar relationship has been established between negative secondary ion yields and electron affinities [4,31,32].

The SIMS "matrix effect" is defined as any change in the secondary ion yields which are caused by changes in the chemical composition or structure of a particular specimen [28]. An example of the SIMS matrix effect is illustrated in Figure 2 for

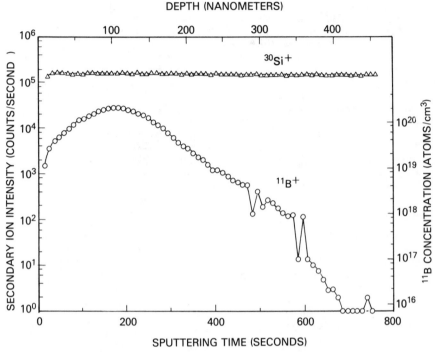

Figure 1.   SIMS depth profile of a $^{11}B^+$ ions implanted Si specimen.

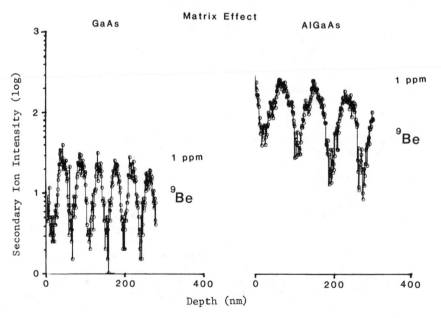

Figure 2.   SIMS depth profile of $^9Be^+$ ions in GaAs and AlGaAs matrices.

Be layers grown by molecular beam epitaxy (MBE) in GaAs and
AlGaAs matrices. Even though the concentration of Be in both
matrices is approximately the same, i.e., 1 ppm or~ 2×10^{16}
atoms/cm^3, the secondary ion intensity of $^9Be^+$ ions differs by
about a factor of ten. The Be$^+$ ion yield from AlGaAs is higher
than from GaAs. The variation of secondary ion yields by an
order of magnitude occurs for most elements implanted into
different semiconductor matrices [33,34].

Other examples of SIMS matrix effects follow. When argon
was used as the primary ion, Slodzian [35] determined that the
ion yield of Ta$^+$ from Ta metal was three orders of magnitude
lower then the ion yield from Ta oxide. Ganjei, Leta, and
Morrison [36] have determined that secondary ion yields do not
change uniformly for different elements in steels and other
metallurgical materials when the residual oxygen pressure of the
sample chamber is increased (a comparison of secondary ion
yields from amorphous and polycrystalline iron based alloys [37]
concluded that a small but significant matrix effect existed for
the matrix elements iron and boron in a series of specimens.)
On the other hand, in glass matrices, Ganjei and Morrison [37]
found that oxygen did not affect the secondary ion yields
probably because the glass matrices were already saturated with
oxygen.

The sputtering yield of a material is also matrix dependent. The sputtering yield is defined as the ratio of the
number of ions and atoms sputtered from the surface of a
specimen to the number of primary ions incident upon the
specimen [28]. Single crystal materials with the same crystal
orientation under the same experimental conditions will have
uniform sputtering rates. The sputtering rate can be determined
from the area that was sputtered and the number of primary ions
that impinged upon that area. However, changes in the crystal
orientation of a single crystal have changed sputtering yields
by up to 50% [33,34]. Also, different chemical phases usually
have different sputtering yields. Sputtering yield effects are
usually considered to be matrix effects.

In particular, Katz et al. [39] found a linear correlation
between sample sputtering yield and average sample mass. With
this relationship the authors could predict ion yields and
detection limits of related compounds. In another study
involving pure elemental matrices, an inverse relationship was
determined between a matrix element's secondary ion yield and
the sputtering rate of that matrix when either cesium or oxygen
was used as the primary ion [40-43]. A similar relationship was
found for secondary ion yields of trace elements in single
element matrices [32]. These authors concluded that only the
ionization potential or electron affinity of the sputtered atom
and the near-surface concentration of oxygen or cesium controlled secondary ion yields. Furthermore, they concluded, that
at least for metal silicides, group IV elements, and GaAs, the
matrix effect is merely an artifact caused by different

sputtering yields. However, Wittmaack [44] has noted that the
precision of the results did not in his opinion justify their
conclusions, and that matrix effects are actually present within
the experimental error of the data which had been presented
logarithmically. Also, important physical processes such as
differential sputtering were not considered, and would preclude
the application of this model to materials with complex
matrices.

Galuska and Morrison [45] showed that the secondary ion
yields and sputtering yields of $Al_xGa_{1-x}As$ matrices are linearly
dependent on the sample composition. Relative ion yield, and
relative sputtering yield calibration lines are used to
determine the concentration of B implanted into a multilayer-
multimatrix specimen. In other work [46], they applied a
point-by-point matrix effect calibration procedure to a variety
of $Al_xGa_{1-x}As$ multilayer-multimatrix specimens grown by MBE.
The procedure used the linear dependence of secondary ion yields
and sputtering yields on matrix composition to quantify depth
profiles through matrix gradients and interfaces. In the actual
procedure, a computer program is used to correct depth profiles
for matrix effects. For example, for $Al_xGa_{1-x}As$ matrices, the
matrix composition and depth at each point of a depth profile is
determined by an iteractive process involving calibration lines
for both relative sputtering yield and the relative ion yield of
$^{75}As^+$. Dopant profiles are then corrected for matrix changes by
use of the appropriate dopant calibration lines.

Rudat and Morrison [47] have determined that instrumental
transmission factors change as a function of the residual oxygen
pressure in the sample chamber for the analyses of single
elemental matrices.

Useful Ion Yields. Quantitation of SIMS results requires one to
determine the relationship between the measured secondary ion
intensities, I, and the original elemental concentration of the
solid. For a particular element M the measured secondary ion
intensity has been defined by Morrison and Slodzian [6] as

$$I = \tau^{\pm} C_m S J_p A \qquad (1)$$

where τ is the practical or useful ion yield, C_m is the concentration of the element M corrected for its isotopic abundance, S
is the total sputtering yield, J_p is the primary beam current
density, and A is the analysis area of the target surface. The
useful ion yield is defined as the ratio of the number of M^+
ions detected to the total number of atoms sputtered from the
solid surface. The difficulty of solving Equation 1 for C_m is
that for complex matrices the useful ion yield and the total
sputtering yield are not known.

An element's useful ion yield is inversely proportional to
the element's first ionization potential and affected by various
instrumental factors and sampling conditions, the chemical state

of the element, and the chemical and physical properties of its
location in the solid. If the sputtering rate can be determined, the chemical composition and density of the matrix are
known, and there is no preferential sputtering, then the useful
ion yield can be determined experimentally. If the matrix of a
standard and a sample are identical, then the measured useful
ion yield from the standard can be used in Equation 1 to solve
for C_m in the specimen. Useful ion yields have been determined
for elements implanted in single crystal semiconductors by Leta
and Morrison [33] and can be used as standards for similar
semiconductor materials. Because instrumental and sampling
parameters affect the useful ion yield, it is necessary to
determine useful ion yield values for the standards at the same
time that the specimen is being analyzed. Depending upon the
matrix in which they are implanted, some of the element's useful
ion yields were found to vary by up to an order of magnitude.

Physical Models. Two basic approaches are used to quantify
secondary ion intensities: physical models and empirical
methods. The physical models consist of several theoretical or
semi-empirical treatments developed to simulate secondary ion
emission [3,48-50]. Although several models have been developed
(see Werner [3] for a recent review) and continue to be applied,
the use of calibration standards (empirical methods) consist-
ently give better results, e.g., accuracies of a factor of 2-3
for physical models compared to 10-20% for empirical methods.

The physical models attempt to account for the influence of
such parameters as ionization potential, work function, and
binding energy on the secondary ion yield in order to describe
the ion emission process. While there are perhaps over 20
different models, some of the well-known models include the
kinetic model [51] which describes ions formed by an Auger
de-excitation process of excited neutral particles emitted by
collision cascade; the auto-ionization model [52-54], valid for
rare-gas ion bombardment, which postulates bombardment induced
inner shell excitation into an auto-ionizing state and relaxation via an Auger process; surface effects models [55-59]
where a particle while being ejected is assumed to change its
electronic structure at or near the surface of the specimen via
electron transitions; and various thermodynamics models including the well-known local thermal equilibrium (LTE) model
[31], simplified 2-parameter [60-61] and 1-parameter [62-65]
LTE models, and a local thermal nonequilibrium model [55-68].

The LTE model has been a point of controversy for several
years [50] because the values used in the Saha-Egget equation
to approximate the temperature and electron density of the
assumed plasma are unrealistically high. The model nevertheless continues to be used in SIMS and for some cases the
results can be quite good. It is recommended, however, that the
model be tested against suitable calibration standards prior to
analyzing unknowns [69-71].

Newbury [72] found that a major source of error in the LTE model originates from instrumental discrimination. Sources of such discrimination include the energy bandpass of the secondary ion spectrometer and conversion efficiencies of the ion detector.

Empirical Methods. The empirical methods use calibration standards to derive sensitivity factors that can be used to determine the unknown concentration of given elements in similar matrices [3]. The sensitivity factors are derived from calibration curves that plot measured secondary ion intensities versus the known concentration of standards. Three types of sensitivity factors have been used: the absolute sensitivity factor, the relative sensitivity factor, and the indexed relative elemental sensitivity factor.

The absolute sensitivity factor (ASF) is defined as

$$ASF_X = \frac{I_X}{C_X f_X} \qquad (2)$$

where I_X is the measured secondary ion intensity of one isotope of element X, C_X is the concentration of element X, and f_X is the isotopic abundances of the elemental isotope being measured.

Quantitative information is obtained by determining the ASF_X from the calibration curve generated using external standards. Ideally, the plot of I_X versus C_X gives a straight lines with slope equal to ASF_X. For best results, the matrix of the specimen must be the same as the external standard. In addition, instrumental factors (such as angle of incidence, collection/detection efficiency, etc) and matrix-dependent factors (sputtering yields, ion yield, etc.) must also be the same. There are no corrections for experimental variations or preferential sputtering. For identical inorganic specimens and standards quantitation with high accuracy has been achieved by the ASF method.

The relative sensitivity factor (RSF) is defined as

$$RSF_{X/ref} = \frac{I_X/C_X f_X}{I_{ref}/C_{ref} f_{ref}} \qquad (3)$$

where I is the measured secondary ion intensity, C is the elemental concentration, f is the isotopic abundance, and the subscripts x and ref denote the analyte and reference elements, respectively.

Quantitative information is obtained by determining the RSF from a calibration curve generated by plotting I_x/I_{ref} versus C_x/C_{ref}. An <u>internal</u> reference element ("ref") is used to eliminate long-term instrument instabilities such as drift in the instrument or fluctuations in the primary ion current. The internal reference element is usually a matrix element. For example, a calibration curve used in our laboratory for the quantitation of Be in GaAs is shown in Figure 3. The secondary ion intensity of $^9Be^+$ (normalized to the internal reference element, the matrix element $^{75}As^+$) is plotted against the concentration of Be in the standards. Since the Be concentration is several orders of magnitude lower than the matrix elements, the concentration of the ^{75}As matrix element, i.e., As, is assumed to be constant.

The reference element included in the RSF method compensates for some types of experimental fluctuations. The RSF method has been applied to the elemental determinations in groups of similar inorganic matrices including glasses [37,73-75], metals [37,75], and minerals [76]. If the specimen and reference matrices are similar, the RSF method has been shown to give a precision of ±15% [73].

The indexed relative elemental sensitivity factor or matrix ion species ratio (MISR) [2,36,77] indexes changes in the relative sensitivity factors to changes in the surface composition, particularly, if the surface is exposed to or bombarded with oxygen. For example, when oxygen is leaked into the specimen chamber, a calibration curve can be generated that plots the RSF's as a function of the MO^+/M^+ ion ratio. The MO^+/M^+ ion ratio is chosen as an internal indicator because it was found to be very sensitive to the oxygen surface coverage which, in turn, affects the secondary ion yields. With the indexed RSF method a substantial improvement in precision to ±5-10% has been achieved over the simple RSF method [2], particularly, for the quantitation of metals.

One problem in quantitative SIMS is caused by variations of the instruments. To evaluate these instrumental factors, Newbury [70] conducted comparative SIMS studies of selected glasses and steels with laboratories in the U.S., Japan, and Europe. He had each laboratory calculate relative sensitivity factors for several elements under a variety of experimental conditions. These results were astonishing and showed that a given relative sensitivity factor varied from 5 to 60. He also compared the measured concentrations with predicted values from physical models.

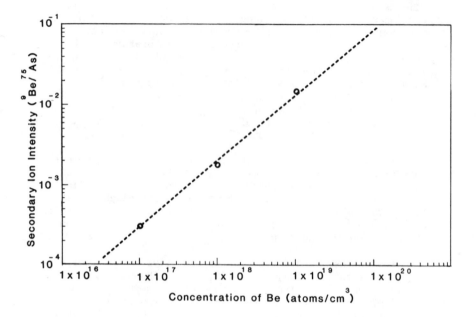

Figure 3. SIMS calibration curve for the quantitation of Be
in GaAs.

Calibration Standards. The empirical methods use calibration standards which are typically from glasses or iron alloys that are chemically doped with elements of known concentrations. The National Bureau of Standards (NBS) supplies several characterized standards of this type.

Metallic glasses are also being used as standards. Rudenauer and coworkers [78, 79] use metallic glasses as standards for several reasons: 1) the metallic glasses are single phase systems and are homogeneous at the micron scale; 2) metallic glasses can be prepared in a broader concentration range than alloying components, and even insoluble elements can form an amorphous phase under suitable conditions; 3) the ion yields of elements in isotopic amorphous alloys are not-dependent on the orientation of the bombardment surface; and 4) amorphous alloys have metallic character.

One problem, however, associated with the use of bulk standards in SIMS is homogeneity. This is particularly acute for the iron alloys. A recent SIMS study concluded that some of the NBS steel standards are unsuitable as standards in SIMS because all of the elements are not homogeneous within the sampling area of 100Å [36, 69]. Therefore, researchers must be careful in both their choice of standard and which elements in each standard are suitable calibrants.

One way to overcome the sample homogeneity problem has been to develop standards by ion implantation [80-83]. Here, the concentration and distribution of the dopants can be controlled more accurately, thereby forming standards with better homogeneity. However, the results with semiconductors, have been much more reliable than with ion-implanted metal standards [82, 83].

Leta and Morrison [82, 83] have described a new empirical method for quantitative SIMS analysis. They use the method of solid-state addition in which they implant their specimens with a known concentration of the element of interest. Since the depth profile of the implanted species has a characteristic Gaussian shape, it is easily distinguished from the element originally present in the specimen. Therefore, the known concentration of the implanted element is used as an internal standard to determine the concentration of the unknown.

Depth Profiling. As the sputtering process in SIMS removes successive atomic layers from the solid, the in-depth composition and distribution of elements can be determined by recording the secondary ion intensity for each element as a function of sputtering time, producing a sputter depth profile. The secondary ion intensities are usually converted to their respective atomic concentrations by the appropriate quantitative methods discussed above and the time axis is converted to a depth (distance) scale (see Figure 1). Although sputter depth profiling has found widespread application in the semiconductor

industry, there are various effects associated with ion-solid
interactions that are not yet well enough understood to convert
the measured depth profile to the true concentration profile of
the specimen.

Many of these effects in SIMS depth profiling have been
reviewed recently by Wittmaack [27], Hofmann [25, 26] and Magee
and Honig [24]. Wittmaack [27] has written an excellent review
detailing the recent progress and basic physical problems
encountered in depth profiling by SIMS. He shows that SIMS has
reached a level of perfection which is unparallel by other
analytical techniques, particularly the in-depth analysis of low
concentration implants. There are analysis problems, however,
caused by the interaction of primary ions with the residual gas,
the adsorption and incorporation of residual gases, sputtering
yield variations due to the accumulation of primary atoms in the
specimen, mass interference between polyatomic ions and the
atomic species under study, and beam-induced relocation of
dopant atoms, i.e.,atomic mixing effects. Hofmann [26] also
summarizes the distorting effects in sputter depth profiling as
due to instrumental factors (e.g., adsorption of residual gases,
redeposition of sputtered species, crater edge effects,
impurities in the ion beam, neutrals from the ion source,
nonuniform ion beam intensity, mean information depth), specimen
characteristics (e.g., original surface roughness, crystalline
and defect structure, multiphases or compounds, insulators), and
radiation induced effects (e.g., primary ion implantation,
atomic mixing, enhanced diffusion and segregation, sputtering
induced roughness, preferential sputtering, and specimen
decomposition).

Hofmann (25, 26) discusses ways to minimize these effects
as well as their influence on depth resolution. For example,
the effects of surface contamination are reduced by using a high
purity ion beam and UHV conditions. Sputtering with reactive
ions reduces topographical change and favors quantitative analysis. Using two ion guns at different angles and rotating the
specimen will suppress incident angle dependent texturing and
core formation. Recoil implantation and atomic mixing effects
are reduced by using heavy primary ions with low energy (<1 keV)
and at oblique incidence. Ion beam rastering provides uniform
sputtering and raster gating avoids crater edge effects. Optimal
results are also expected for solids having a) flat and smooth
surfaces, b) amorphous structures with no second phases, and c)
components with similar sputtering yields. Specimens with high
electrical and thermal conductivity are also preferred.

In another review, Magee and Honig [24] discuss three
important aspects of depth profiling by SIMS: depth resolution,
dynamic range and sensitivity. First, the depth resolution is a
measure of the profile quality. They point out that the depth
resolution is limited by atomic mixing effects and the flatness
of the sputtered crater within the analyzed area. Second, the
dynamic range of depth profiles is limited by crater edge

effects, neutral beam effects, spectral interferences, residual gas contamination, recontamination from previously sputtered materials, and noise from the detection of nonfilterable particles. Third, in the absence of a measurable background, the ultimate sensitivity of a depth profile analysis is dependent on ion yield and analyzed area and can be increased only at the expense of depth and spatial resolution.

Organic SIMS

SIMS has become a diverse tool in the study of many different substances other than metals and semiconductors. This part of the paper discusses the secondary ion emission of molecular and polyatomic ions from the surfaces of organic compounds including polymers and biomolecules.

Ionization Methods/Processes. The recent development of several new ionization methods in mass spectrometry has significantly improved the capability for the analysis of nonvolatile and thermally labile molecules [18-23]. Several of these methods (e.g., field desorption (FD), Californium-252 plasma desorption (PD), fast heavy ion induced desorption (FHIID), laser-desorption (LD), SIMS, and fast atom bombardment (FAB) or liquid SIMS) desorb and ionize molecules directly from the solid state, thereby reducing the chance of thermal degradation. Although these methods employ fundamentally different excitation sources, similarities in their mass spectra, such as, the appearance of protonated, deprotonated, and/or cationized molecular ions, suggest a related ionization process.

The ionization process in SIMS is undoubtedly dependent on such physical properties as the ionization potential or the electron affinity of a given species. Researchers in molecular SIMS have defined phenomenologically three distinct ionization processes based on the type of ions created and their relative ease of ionization [18].

The first process involves electron ionization to form radical $M^{+\cdot}$ molecular ions. This process has been observed primarily for nonpolar molecules. The proposed mechanisms are charge-exchange transitions between sputtered ions and the neutral organic molecules or electron attachment of low-energy secondary electrons to neutral molecules. The fragmentation reactions of the $M^{+\cdot}$ ions usually follow the dissociation pathways for odd-electron gas-phase ions.

The second process involves the formation of protonated or cationized molecular ions, i.e., $[M+H]^+$ or $[M+C]^+$, where the cationizing species C is usually a metal ion from the substrate, matrix, or an impurity. The basic fragmentation process

involves the loss of neutral molecules from the even-electron
ions due to unimolecular dissociation reactions which are common
in other forms of mass spectrometry, e.g., chemical ionization
mass spectrometry.

The third process of ionization in molecular SIMS involves
the direct emission of intact (or performed) charged species
from the solid state as [M-anion]$^+$ or [M-cation]$^-$ ions. SIMS
studies of organic salts yield intense cationic and anionic
species with little fragmentation [85]. The secondary ion
intensity for the organic salts is generally two orders of
magnitude higher than that observed in cationization. In fact,
the relative ionization efficiency (number of secondary ions
desorbed/number of molecules desorbed) for these processes in
molecular SIMS is direct emission > cationization > electron
ionization. The higher efficiency of the direct emission
process lowers the detection limits for organic salts in SIMS
such that picogram quantities can be detected [86].

Sample Preparation. The methods of sample preparation affect
the chemical and physical properties of the sample molecules and
hence can profoundly influence the secondary ion formation/
emission process. In earlier molecular SIMS studies the samples
were prepared by placing a dilute solution of the compound onto
an acid-etched Ag foil [87, 88]. The acid etched surface
provides, a substrate onto which thin layers of the compound can
be deposited from solutions with extended concentration ranges.
If on the other hand, the substrate was not etched and the
concentration of the solution was too high, the adsorbed
molecular film would grow too thick and consequently quench the
secondary ion emission.

Derivatizing the neutral sample molecules to form ionic
species enhances their secondary ion yield. Derivatization of
the neutral sample molecules can often be accomplished by simply
adding acid or base to the sample solution, or through the
chemical modification of specific functional groups of the
molecules, e.g., quaternization reactions [20, 89].

Several other sample preparation methods were developed to
simplify the solution-deposition procedures. For example, Cooks
and coworkers studied adduct ion formation (cationization) of
several organic compounds when the organic was burnished
(rubbed) onto a metal foil, mixed with a metal salt and then
burnished onto a metal foil, or just mixed with metal powders or
salts and pressed into pellets. Not only did the SIMS spectra
show dramatic differences is the efficiency of adduct ion
formation for different metals, but the sample preparation
methods had an equally dramatic effect on the SIMS spectra [18].

Matrix assisted SIMS. Molecular ion emission from a number of
solid-state and liquid matrices has been investigated recently.
There are two types of solid-state matrices used in molecular
SIMS namely, low temperature matrices (such as the rare-gas

solids [90] and frozen molecular solids [91-95]) and room
temperature matrices (such as ammonium chloride, NH_4Cl [96-98]
and carbon [99, 100]). One important property of these
matrices is their ability to matrix isolate and dilute the
sample molecules. In addition, because NH_4Cl is not chemically
inert, it can protonate molecules more basic than ammonia [96]
enhancing their ion yield [97, 98]. The carbon matrix, on the
other hand, has the unique property of being able to strongly
adsorb many organic molecules, even volatile and nonpolar
compounds [99, 100].

The recent use of liquid matrices in SIMS has led to
several significant accomplishments, particularly for the
analysis of biomolecules by FAB mass spectrometry [101]. In the
FAB/liquid SIMS technique, the analyte is dissolved in a liquid
matrix, such as glycerol, at some optimum concentration in order
to provide a surface which, during particle bombardment, is
constantly replenished with sample molecules that diffuse from
the bulk to the surface. The mobility of the sample molecules
in the liquid matrix is an important property of this type of
matrix.

We have recently investigated another type of mobile matrix
- a liquid metal [100, 102]. Here, we discovered that ion
bombardment of the liquid metal surface, upon which sample
particles were deposited, resulted in movement of the sample
species towards the primary ion beam where they were desorbed
and finally detected by the mass analyzer.

Carbon. Since activated carbons or charcoals are used as
adsorbents for airborne or aqueous pollutants, one objective of
a recent investigation was to develop a new analytical method
for the direct and rapid identification of organic compounds
adsorbed on carbon. One important class of organic compounds -
the polycyclic aromatic compounds (PACs) were previously
difficult to analyze by SIMS because of their low ion yield and
low adsorption energies on metal surfaces. By adsorption on
carbon volatile organic compounds such as the substituted
benzenes (toluene, xylene, and mesitylene) are readily detected
without cryogenically cooling the matrix [99]. In fact, the
molecular ion signal of toluene lasted for 1/2 to ~1 hour when
the carbon was saturated with toluene. Since toluene was
detected, but not benzene, the minimum adsorption energy
necessary to permit detection is taken to be the heat of
adsorption for toluene or ~11 kcal/mol.

When the larger nonvolatile PACs are adsorbed on carbon,
molecular and cationized ions are also detected readily (99).
We believe that the high surface area and porosity of the carbon
provides a three-dimensional matrix over which the sample
molecules are dispersed and from which ions can be emitted for
extended periods of time. For example, 2 µg of a PAC on carbon

emitted cationized ions for over one hour. A detection limit of
~ 1 ng of phenanthrene on carbon was measured under dynamic
sputtering conditions (i.e., > 1 X 10^{-7} A/cm^2).

        Figures 4 and 5 show the molecular SIMS spectra of phenan-
threne and 9-aminophenanthrene taken from four different
matrices: Ag foil, NH$_4$Cl, carbon, and a liquid metal [103]. We
found that the molecular ion yields and ionization efficiencies
for phenanthrene (at m/z 178) were similar from the Ag, NH$_4$Cl
and carbon matrices, but that the signal-to-background ratio was
much better from the carbon matrix. (The mass peak at m/z 191
is attributed to a contaminant in either the vacuum chamber or
sample. Since many different polycyclic aromatic hydrocarbon
compounds had been studied over many months, the vacuum chamber
and sample carrousel had to be cleaned periodically to reduce a
contamination problem.) For the 9-aminophenanthrene, on the
other hand, its secondary ion yield and ionization efficiency
were enhanced by the NH$_4$Cl matrix as evident by the higher
counting rate in Figure 5. This observation demonstrates how
the chemical properties of NH$_4$Cl can influence the secondary ion
emission of molecules containing certain functional groups.
Further comparison of the matrices showed that the Ag matrix was
the easiest to prepare but required a static primary beam to
ensure low background and good signal-to-noise. The NH$_4$Cl
matrix enhanced the ion emission for most substituted compounds
and worked best (i.e., high signal-to-noise) with a static ion
beam. The unsubstituted PACs were most easily analyzed from
carbon where use of a dynamic primary beam was possible without
increasing the background ion intensity.

Liquid Metal Substrate. Initial experiments using a CAMECA ion
microscope (5.5 keV Ar$^+$ or O$_2^+$ ion beam at >1 X 10^{-6} A/cm^2)
demonstrated that the liquid metal (a gallium/indium alloy)
provided a suitable substrate from which long-lived M$^{+\cdot}$ ion
emission of organic molecules occurred while using a dynamic
primary ion beam [100, 102]. The experimental set-up for the
liquid metal substrate is shown schematically in Figure 6. We
found that ion bombardment of the liquid metal surface, upon
which sample particles were deposited, resulted in movement of
the sample species towards the primary ion beam where they are
desorbed and finally detected by the mass analyzer. This liquid
metal substrate offers several advantages over conventional
liquid or solid matrices (see Table I). For example, the
gallium/indium alloy has a smooth surface onto which solids can
be dispersed. The liquid metal is also conducting and has a low
vapor pressure.

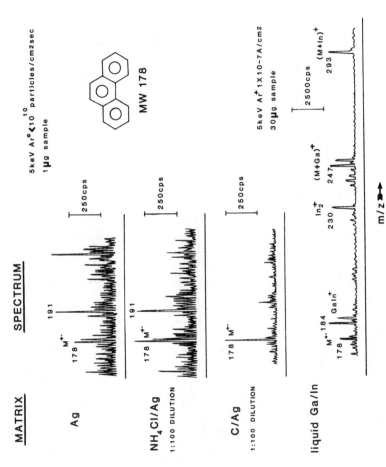

Figure 4. SIMS spectra of phenanthrene taken from silver foil, ammonium chloride, carbon, and liquid gallium/indium alloy. Reproduced with permission from Ref. 103. Copyright 1983, Elsevier Science Publishing Co.

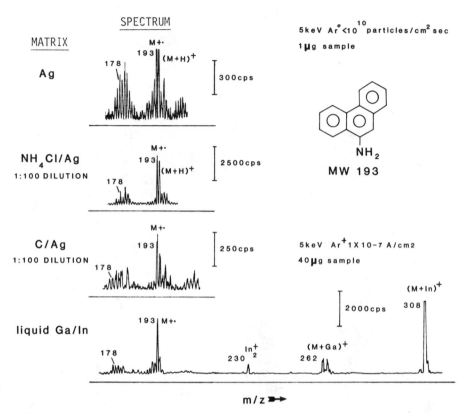

Figure 5.   SIMS spectra of 9-aminophenanthrene taken from silver foil, ammonium chloride, carbon, and gallium/indium alloy.   Reproduced with permission from Ref. 103.   Copyright 1983, Elsevier Science Publishing Co.

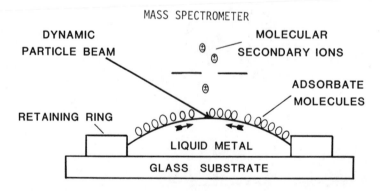

Figure 6.   Schematic diagram of the experimental set-up for dynamic SIMS analysis using a liquid metal substrate.   Reproduced with permission from Ref. 103.   Copyright 1983, Elsevier Science Pub. Co.

TABLE I. Comparisons of the Liquid Metal and Glycerol Substrates for Molecular SIMS

Property	Matrix	
	Liquid Metal[a]	Organic[b]
Sample Introduction	Surface	Bulk Solution
Interference Peaks	Few	Many
Physiochemical Nature of Surface/Liquid	Smooth/Inert	Polar/Selective Solvent
Vapor Pressure	10^{-11} torr @ 435°C	3×10^{-4} torr @ 30°C
Required Sample Size	1-50 g	10-500 ng
Conductivity	High	Low (must add salt if charging

[a] Ga/In alloy
[b] Glycerol

SIMS spectra of phenanthrene and 9-aminophenanthrene analyzed from the liquid metal surface are shown in Figures 4 and 5. The liquid metal substrate requires a dynamic ion beam and sample concentration of 10-40 g for optimum performance in order to sustain the long-lived molecular ion emission. Cationization of the organic molecules by Ga^+ and In^+, i.e., $[M +Ga]^+$ and $[M + In]^+$, occurs with high efficiency.

Polymers. Molecular SIMS is an especially attractive surface analytical technique for the characterization of polymer surfaces because of its high surface sensitivity, molecular specificity over an extended mass range, and versatility in ionization. However, only a few polymer SIMS studies have been reported [104-113] due to the serious sample charging problems that occur with thick insulating films.

A number of experimental methods have been used to reduce the charging of insulator surfaces [114, 115] especially in AES or SIMS experiments where charged primary beams are employed. These methods include (a) electrical compensation, (b) flooding with low-energy electrons or ions, (c) evaporating conducting films or grids onto the surface, (d) increasing the partial pressures of oxygen in order to increase the surface conductivity, (e) emitting secondary electrons from a metal foil placed near the insulator surface, (f) using negative primary ions, (g) using neutral primary particles, (h) using cesium primary ions

or cesium surface overlayers, and (i) heating the specimen.
Although a number of these methods have been highly successful
in specific cases, the majority of them are experimentally
cumbersome and irreproducible.

To overcome this problem, we have modified a commercial ion
gun to generate a diffuse fast-atom beam [116, 117]. The ion
beam neutralizer shown in Figure 7 consists of a multi-hole
metal plate through which the primary ions pass. The ions are
neutralized by the ion/surface interactions that occur as the
beam passes through the metal aperatures and by charge-exchange
reactions that occur within the gun assembly. A repeller grid
is used to remove the residual ions from the neutralized beam.

The SIMS spectrum of a 0.5 mm thick film of polystyrene
cast on silver is shown in Figure 8. The characteristic ions,
phenyl-type (m/z 77, 78, 79), benzyl (m/z 91), and protonated
styrene (m/z 105) along with higher m/z ions resembling those
from pyrolysis studies, are observed.

Derivatizaion SIMS. The sensitivity of SIMS for quaternary
ammonium salts extends to the nanogram or subnanogram range
[86]. This enhanced sensitivity forms the basis of a new
analytical method to detect organic compounds in mixtures.
Specifically, a target compound in a mixture is selectively
derivatized to form a quaternary ammonium salt and the deriv-
ative is analyzed using SIMS. The compound is derivatized
directly in the mixture without sample extraction or clean-up
prior to analysis. In addition, because only the derivatized
species are detected with high sensitivity, the mass spectra
are much simpler and compounds in mixtures can be analyzed using
much less complex instrumentation. To date, we have applied
these methods to analyze drug compounds from urine [18], to
detect aldehydes and ketones [119] in air or adsorbed on
activated charcoals [120], to sequence peptides [121] and other
biomolecules [122].

Detection of Drugs. SIMS has been applied previously to the
detection of several drugs in urine [123]. We have demonstrated
picogram sensitivity for drugs in urine by first performing a
simple derivatization with methyl iodide, according to Equation
4 to form a quaternary ammonium salt of the drug prior to
analyzing the urine with SIMS [118].

BASIC DRUG* DERIVATIZATION SCHEME

Basic Drug (in water or urine)/ MeOH + CH$_3$I

$$\xrightarrow[\substack{60°C \\ 30 \text{ min.}}]{K_2CO_3}$$ **Methylated Quaternary Salt**   (4)

*Drugs containing amines

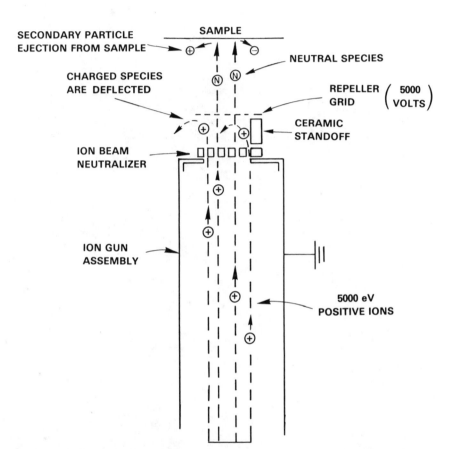

Figure 7. Schematic diagram of commercial ion gun modified with an ion beam neutralizer. Reproduced with permission from Ref. 116. Copyright 1983, Elsevier Science Publishing Co.

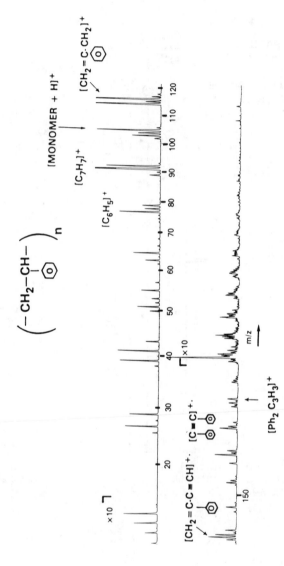

Figure 8. SIMS spectrum of a 0.5 mm thick insulating film of polystyrene cast on silver. Reproduced with permission from Ref. 116. Copyright 1983, Elsevier Science Publishing Co.

Figure 9 shows a commercial drug mixture in human urine underivatized and derivatized with methyl iodide.   In Figure 9a the propoxyphene and methamphetamine are not observed from the urine.   However, upon derivatization, molecular ions from both can be seen and the ion intensity from methadone and quinine have been increased by a factor of ten.

We have derivatized and observed other drugs such as morphine, codeine, amphetamine, and meperidine with similar results.   Cocaine could not be derivatized under the conditions that we employed and was only detected as the protonated molecular ion.

We have also employed other derivatization techniques to selectively label other moieties such as carbonyls and applied these to the selective detection of steroids [119].   Girard's Reagent P reacts with the carbonyl-containing compounds according to Equation 5.

| Girard's Reagent P | Aldehyde/Ketone | Derivative | |
| (152) | (M) | (M+134) | (5) |

The ions corresponding to the derivatized carbonyl-containing compounds are detected as $[M + 134]^+$, where M is the molecular weight of the aldehyde or ketone.

Steroids such as progesterone, testosterone, and cortisone have been derivatized and analyzed with this method.   Cortisone was chosen to demonstrate the enhanced sensitivity obtained with the derivatization/SIMS technique.   The SIMS results for underivatized and derivatized cortisone are presented in Figure 10.   The ions that are observed from cortisone include the molecular ion $[M + H]^+$ (m/z 361), the silver cationized ion $[M + Ag]^+$ (m/z 467 and 469), and the fragment ion at m/z 407 and 409 corresponding to the loss of the $OCHCH_2OH$ group.

With 100 ng of cortisone only $[M + Ag]^+$ is observed.   The SIMS spectra of derivatized cortisone, on the other hand, show a single peak corresponding to the molecular ion of the derivatized species and an ion of low abundance corresponding to the same fragmentation observed for the underivatized species.   The derivatized cortisone was detected at 10 ng with the same signal-to-noise ratio as that for 100 ng of underivatized cortisone, which is approximately an order of magnitude improvement in the detection sensitivity.

Quantitative Analysis.   The derivatization/SIMS method can also be applied quantitatively.   For example, two ketones, acetone and d_6-acetone, were mixed in various concentrations in the range of 0.1 to 10 mg/mL and in the ratios of d_0/d_6 = 10/1, 1/1,

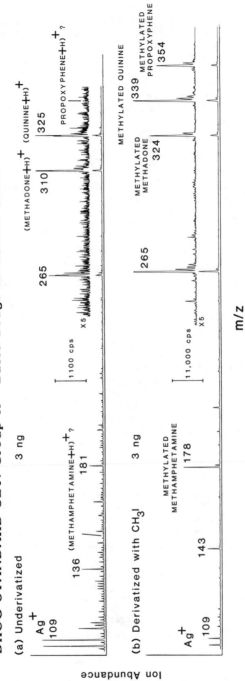

Figure 9. SIMS spectra of an equal mixture of methamphetamine, methadone, quinine, and propoxyphene in human urine. Reproduced with permission from Ref. 118. Copyright 1985, Heyden & Son Ltd. (London).

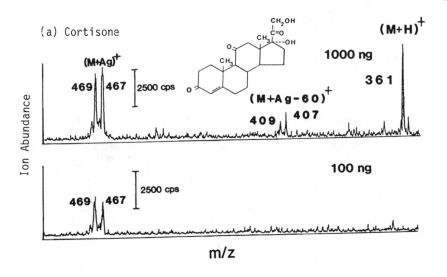

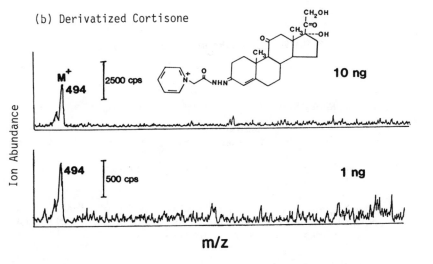

Figure 10. SIMS spectra of a) underivatized cortisone and b) derivatized cortisone.

1/10. Derivatizing the mixture using the Girard's reagent
reaction given by Equation 5 produces the results presented in
Figure 11 [119]).
   It is evident from the spectra that absolute ion abundances
cannot be used for quantitation. However, the ion abundance
ratio of the two compounds remained constant over the selected
concentration range. Therefore, internal standards must be used
for quantitation by SIMS. This has been found to be true for
fast-atom bombardment mass spectrometry (FABMS), also [124].

Sequencing of Biomolecules. The enhanced detection of charged
over uncharged compounds also forms the basis for a method of
sequencing of biomolecules [121]. A peptide is selectively
labeled at the N- or C-terminus with a charged group (Equation
6), partially cleaved, esterified and acylated, and the mixture
examined by SIMS. Only the charged, labeled components in the
mixture are observed. Since the charge is due to the label and
the label is at only one end, the sequence of the peptide can be
readily reconstructed. The spectra of two peptides sequenced
from the N-terminus are shown in Figure 12. Only low nanogram
levels of each charged species are necessary for detection. It
should also be emphasized that the SIMS spectra were taken of
the unpurified mixtures.

$$H_2NGlyGlyPheOH \xrightarrow{\text{MeI}} Me_3\overset{+}{N}GlyGlyPheOMe \qquad (6)$$

$$\xrightarrow{\text{Hydrolysis}} \xrightarrow[\text{HCl}]{\text{MeOH}} \quad F_3C \overset{O}{\underset{2}{\diagup}} O \longrightarrow$$

m/z				
336	$Me_3\overset{+}{N}GlyGlyPheOMe$		$F_3C\diagdown NPheOMe$	
189	$Me_3\overset{+}{N}GlyGlyOMe$	Charged and observed	$F_3C\diagdown NGlyOMe$	Not charged and not observed
132	$Me_3\overset{+}{N}GlyOMe$		$F_3C\diagdown NGlyPheOMe$	

Reconstructed sequence

   Our method of enhancing the emission of biological
molecules by attaching charged groups has also been applied to
oligosaccharides. Sugars are normally only observed with
difficulty using SIMS. However, reducing sugars can be readily
labeled in aqueous media with Girard's reagents. Once labeled,
the sugars can be observed at much lower concentrations than the
corresponding underivatized materials. Figure 13 shows the SIMS
spectra of underivatized maltose deposited from solution onto a
silver surface (Fig 13a), mixed in an ammonium chloride matrix
-assisted SIMS - (Fig 13b and c), and derivatized maltose
(Fig 13d). The derivatized sugar can be observed at 100 times
lower concentrations than the corresponding underivatized
material.

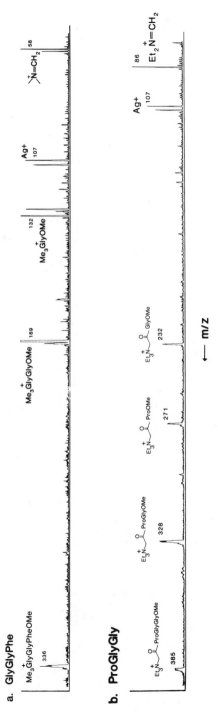

Figure 11.   SIMS spectra of the derivatized a) glycylglycyl-
phenylalanine and b) prolyglycylglycine.   Reproduced from Ref. 121.

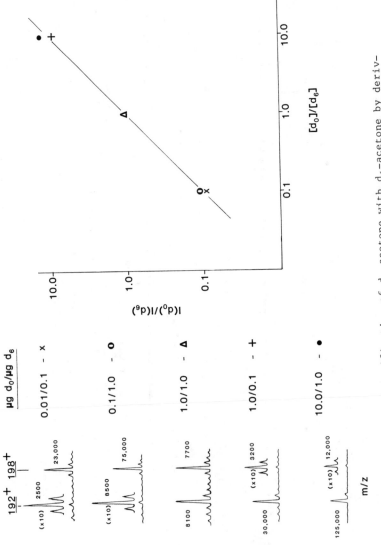

Figure 12. Quantification of d_0-acetone with d_6-acetone by derivatization/SIMS. Reproduced with permission from Ref. 119. Copyright 1985, Elsevier Science Publishing Co.

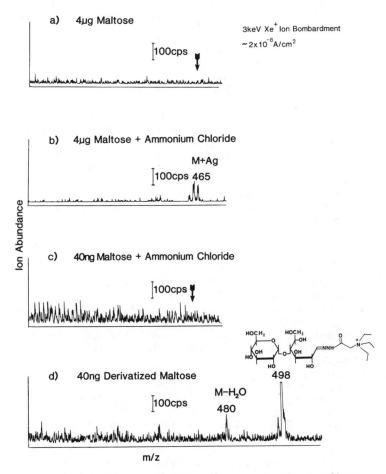

Figure 13. SIMS spectra of maltose and derivatized maltose.

Literature Cited

1. Behrisch, R., Ed. "Sputtering by Particle Bombardment I";
    TOPICS IN APPLIED PHYSICS, Vol. 47, Springer Verlag: New
    York, 1981.
2. McHugh, J. A. In "Methods of Surface Analysis"; A. W.
    Czanderna, Ed.; Elsevier: New York, 1975; p. 223.
3. Werner, H. W. Surf. Interface Anal. 1980, 2, 56-74.
4. Wittmaack, K. Nucl. Instrum. Methods 1980, 168, 343-356.
5. Morrison, G. H. Springer Ver. Chem. Phys. 1982, 19, 244-56.
6. Morrison, G. H. Slodzian,; G. Anal. Chem 1976, 932A-943A.
7. Benninghoven, A. Evans, Jn., C. A.; Powell, R. A.; Shimizu,
    R.; Storms, H. A., Eds.; "Secondary Ion Mass Spectrometry
    SIMS II"; Springer-Verlag: New York, 1979.
8. Benninghoven, A.; Giver, J.; Laszlo, J.; Riedel, M.; Werner,
    H. W., Eds.; "Secondary Ion Mass Spectrometry SIMS III";
    Springer-Verlag: New York, 1982.
9. Benninghoven, A.; Okano, J.; Shimizu, R.; Werner, H. W.,
    Eds.; "Secondary Ion Mass Spectrometry SIMS IV"; Springer-
    Verlag: New York, 1984.
10. Benninghoven, A. Z. Physik 1970, 230, 403.
11. Macfarlane, R. D.; Torgerson, D. F. Science 1976, 191,
    920-925.
12. Benninghoven, A. Surf. Sci. 1971, 28, 541.
13. Benninghoven, A. Surf. Sci. 1973, 35-427.
14. Wittmaack, K. Surf. Sci. 1979, 89, 668.
15. Winograd, N.; Garrison, B. J. Acc. Chem. Res. 1980, 13, 406.
16. Winograd, N. Prog. Solid St. Chem. 1982, 13, 285-375.
17. Garrison, B. J.; Winograd, N. Science 1982, 216, 805-812.
18. Day, R. J.; Unger, S. E.; Cooks, R. G. Anal. Chem. 1980, 52,
    557A-572A.
19. Colton, R. J. J. Vac. Sci. Technol. 1981, 18, 737-747.
20. Busch, K. L.; Cooks, R. G. Science 1982, 218, 247-254.
21. Benninghoven, A., Ed. "Ion Formation From Organic Solids";
    Springer-Verlag: New York, 1983.
22. Sundqvist, B., Ed. "Ion Induced Desorption of Molecules From
    Bioorganic Solids"; Nucl. Instrum. Methods 1982, 198,
    1-174.
23. "Texas Symposium on Particle Induced Desorption Mass
    Spectrometry"; Int. J. Mass Spectrom. 1983, 53, 1-366.
24. Magee, C. W.; Honig, R. E., Surf. Interface Anal. 1982, 4,
    35-41.
25. Hofmann, S. Surf. Interface Anal. 1980, 2, 148-60;
26. Hofmann, S. Springer Ser. Chem. Phys 1982, 19, 186-200.
27. Wittmaack, K. Radiat. Eff. 1982, 63, 205-18.
28. ASTM Standards, E 673, Standard Definitions of Terms
    Relating to Surface Analysis
29. Storms, H. A.; Brown, K. F.; Stein, J. D. Anal. Chem. 1977,
    49, 2023.
30. Liebl, H. J. Vac. Sci. Technol., 1975, 12, 385.
31. Andersen, C. A.; Hinthorne, J. R. Anal. Chem. 1973, 45,
    1421.
32. Deline, V. R.; Katz, W.; Evans, Jr., C. A. Appl. Phys. Lett
    1978, 33, 832.

33. Leta, D. P.; Morrison, G. H. Anal. Chem. 1980, 52, 277-280.
34. Leta, D. P.; Morrison, G. H. Anal. Chem. 1980, 52, 514-519.
35. Slodzian, G. Surf. Sci 1975, 48, 161.
36. Ganjei, J. D.; Leta, D. P.; Morrison, G. H. Anal. Chem 1978, 50, 2034-39.
37. Ganjei, J. D.; Morrison, G. H. Anal. Chem 1978, 50, 2034-39.
38. Riedel, M.; Gnaser, H.; R udennauer, F. G. Anal. Chem. 1982, 54, 291.
39. Katz, W.; Williams, P.; Evans, Jr., C. A. Surf. Interface Anal 1980, 2, 120-121.
40. Deline, V. R. Springer Ser. Chem Phys. 1979, 9, 48-52.
41. Deline, V. R.; Evans, Jr., C. A.; Williams, P. Appl. Phys. Lett. 1978, 33, 578.
42. Williams, P.; Katz, W.; Evans, Jr., C. A. Nucl. Instrum. Methods 1980, 168, 373-7.
43. Williams, P.; Deline, V. R.; Evans, Jr., C. A.; Watz, W. J. Appl. Phys. 1981, 52, 530-2.
44. Wittmaack, K. J. Appl. Phys. 1981, 52, 527-9.
45. Galuska, A. A.; Morrison, G. H. Anal. Chem. 1983, 55, 2051-55.
46. Galuska, A. A.; Morrison, G. H. Anal Chem. 1984, 56, 74-77.
47. Rudat, M. A.; Morrison, G. H. Int. J. Mass Spectrom. Ion Phys. 1979, 30, 233.
48. Blaise, G.; Noutier, A. Surf. Sci. 1979, 90, 495.
49. Williams, P. Surf Sci 1979, 90, 588.
50. Krauss, A.; Krohn, V. E. Mass Spectrom. 1981, 6, 118-52.
51. Joyes, P. J. Phys. (Paris) 1969, 30, 365.
52. Blaise, G.; Slodzian, G. J. Phys. (Paris) 1970, 31, 93.
53. Blaise, G.; Slodzian, G. J. Phys. (Paris) 1974, 35, 237.
54. Blaise, G.; Slodzian, G. J. Phys. (Paris) 1974, 35, 243.
55. Sroubek, Z. Surf. Sci 1974, 44, 47.
56. Schroer, J. M.; Rhodin, T. N.; Bradley, R. C. Surf. Sci. 1973, 34, 571.
57. Gries, W. H.; R udenauer, F. G. Int. J. Mass Spectrom. Ion Phys. 1975, 18, 111.
58. Cini, M. Surf. Sci. 1976, 54, 71.
59. Antal, J. Phys. Lett. A 1976, 55, 281.
60. Simons, D. S.; Baker, J. E.; Evan, Jr., C. A. Anal. Chem. 1976, 48, 1341.
61. R udenauer, F. G.; Steiger, W. Vacuum 1976, 26, 537.
62. R udenauer, F. G.; Steiger, W.; Werner, H. W. Surf Sci, 1976, 54, 553.
63. Morgan, A. E.; Werner, H. W. J. Microsc. Spectros. Electron 1978, 3, 495.
64. Morgan, A. E.; Werner, H. W. J. Chem. Phys. 1978, 68, 3900.
65. Morgan, A. E.; Werner, H. W. Mikrochem Acta II 1978, 31.
66. Jurela, Z. Int. J. Mass Spectrom. Ion Phys. 1981, 37, 67-75.
67. Sroubek, Z. Nucl. Instr. Methods 1982, 194, 533-39.
68. Jurela, Z. Nucl. Instr. Methods 1982, 194, 597-602.
69. Illgen, L.; Mai, H.; Seidenkranz, U.; Voightmann, R. Surf. Interface Anal. 1980, 2, 77-84.
70. Newbury, D. E. Springer Ser. Chem. Phys. 1979, 9, 53-7.

71. Okutani, T.; Shimizu, R. Springer Ser. Chem. Phys. 1979, 9, 7-11.
72. Newbury, D. E. Electron Microsc., Proc. Eur. Cong. 7th 1980, 3, 212-13.
73. Smith, D. H.; Christie, W.H. Int. J. Mass Spectrom Ion Phys 1978, 26, 61.
74. Newbury, D. E. Scanning 1980, 3, 110.
75. Smith, D. H.; Christie, W. H. Int. J. Mass Spectrom. Ion Phys 1978, 26, 61.
76. Havette, A.; Slodzian, G. J. Physique L. 1980, 41, L247.
77. Christie, W. H.; Eby, R. E.; Anderson, R. L.; Kollie, T. G. Appl. Surf. Sci 1979, 3, 329-47.
78. Gnaser, H.; Riedel, M.; Martin, J.; R udenauer, F. G. Springer Ser Chem. Phys 1982, 19, 282-7.
79. Riedel, M.; Gnaser, H.; R udenauer, F. G. Anal. Chem. 1982, 54, 290-4.
80. Gries, W. H. Int. J. Mass Spectrom. Ion Phys. 1979, 30, 97-112.
81. Sigmon, T. W. Springer Ser. Chem. Phys. 1979, 9, 80-4.
82. Leta, D. P.; Morrison, G. H. Springer Ser. Chem. Phys. 1979, 9, 61.
83. Leta, D. P.; Morrison, G. H. Anal. Chem 1980, 52, 277-280.
84. Leta, D. P.; Morrison, G. H. Anal. Chem 1980, 52, 514-519.
85. Liu, L. K.; Unger, S. E.; Cooks, R. G. Tetrahedron 1981, 37, 1067-73.
86. Unger, S. E.; Ryan, T. M.; Cooks, R. G. Anal. Chem. Acta 1980, 118, 169-74.
87. Benninghoven, A.; Jaspers, D.; Sichtermann, W. Appl. Phys. 1976, 11, 35.
88. Colton, R. J.; Murday, J. S.; Wyatt, J. R.; DeCorpo, J. J. Surf. Sci 1979, 84, 235.
89. Busch, K. L.; Unger, S. E.; Vincze, A.; Cooks, R. G.; Keough, T. J. Am. Chem. Soc. 1982, 104, 1507-11.
90. Michl, J. Int. J. Mass Spectrom. Ion Phys. 1983, 53, 255-272.
91. Barber, M.; Vickerman, J. C.; Wolstenholme, J. J. Chem. Soc., Faraday Trans 1980, 76, 549.
92. Lancaster, G. M.; Honda, F.; Fukada, Y.; Rabalais, J. W. J. Am. Chem. Soc. 1979, 101, 1951.
93. Jonkman, H. T.; Michl, J. JCS Chem. Comm 1978, 751.
94. Jonkman, H. T.; Michl, J.; King, R. N.; Andrade, J. D. Anal. Chem. 1978, 50, 2078.
95. Jonkman, H. T.; Michl, J. J. Am. Chem. Soc. 1981, 103, 1007 and 1564.
96. Liu, L. K.; Busch, K. L.; Cooks, R. G. Anal. Chem. 1981, 53, 109.
97. Busch, K. L.; Hsu, B. H.; Xie, Y.-X.; Cooks, R. G. Anal. Chem. 1983, 55, 1157-60.
98. Hsu, B. H.; Xie, Y.-X., Busch, K. L.; Cooks, R. G. Int. J. Mass Spectrom. Ion Phys. 1983, 51, 225-33.
99. Ross, M. M.; Colton, R. J. Anal. Chem 1983, 55, 150-3.
100. Ross, M. M.; Colton, R. J. J. Vac. Sci. Technol. 1983, A1, 443.

101. Barber, M.; Bordoli, R. S.; Elliott, G. J.; Sedgwick, R. D.; Taylor, A. N. Anal. Chem 1982, 54, 645A.
102. Ross, M. M.; Colton, R. J. Anal. Chem 1983, 1170-1.
103. Colton, R. J. Nucl. Instrum. Methods Phys. Res. 1983, 218, 276-286.
104. Werner, H.W. Microchim Acta, Suppl. VII 1977, 63.
105. Gardella, Jr., J.A.; Hercules, D.M., Anal. Chem. 1980, 52, 226.
106. Holm, R.; Storp, S., Surf. Interface Anal. 1980, 2, 109.
107. Gardella, Jr., J.A.; Hercules, D.M., Anal. Chem. 1981, 53, 1879.
108. Campana, J.E.; DeCorpo, J.J.; Colton, R.J., Appl. Surf. Sci. 1981, 8, 337.
109. Campana, J.E.; S.L. Rose, Int. J. Mass Spectrom. Ion Phys. 1983, 46, 483.
110. Briggs, D.; Wootton, A.B., Surf. Interface Anal. 1982, 4, 109.
111. Briggs, D., Surf. Interface Anal. 1982, 4, 151-155.
112. Briggs, D., Surf. Interface Anal. 1983, 5, 113-118.
113. Briggs, D.; Hearn, M.J.; Ratner, B.D., Surf. Interface Anal. 1984, 6, 184-192.
114. Werner, H.W.; Morgan, A.E., J. Appl. Phys. 1976, 47, 1232.
115. Borchardt, G.; Scherrer, H.; Weber, S.; Scherrer, S., Int. J. Mass Spectrom. Ion Phys. 1980, 34, 361.
116. Ross, M.M.; Wyatt, J.R.; Colton, R.J.; Campana, J.E., Int. J. Mass Spectrom. Ion Phys. 1983, 54, 237-247.
117. Ross, M.M.; Colton, R.J.; Rose, S.L.; Wyatt, J.R.; DeCorpo, J.J.; Campana, J.E., J. Vac. Sci. Technol. 1984, A2, 748-750.
118. Kidwell, D.A.; Ross, M.M.; Colton, R.J., Biomed. Mass Spectrom., in press.
119. Ross, M.M.; Kidwell, D.A.; Colton, R.J., Int. J. Mass Spectrom. Ion Phys., 1985, 63, 141-148.
120. Ross, M.M.; Kidwell, D.A.; Campana, J.E., Anal. Chem., 1984, 54, 2142-2145.
121. Kidwell, D.A.; Ross, M.M.; Colton, R.J., J. Am. Chem. Soc., 1984, 106, 2219-20.
122. Kidwell, D.A.; Ross, M.M.; Colton, R.J., Presented at the 32nd Annual Conference on Mass Spectrometry and Allied Topics, San Antonio, TX, 1984, paper no. WOC2.
123. Sichtermann, W.; Junack, M.; Eicke, A.; Benninghoven, A., Fresenius Z. Anal. Chem., 1980, 301, 115.

RECEIVED April 16, 1985

Fast Atom Bombardment Combined with Tandem Mass Spectrometry for the Study of Collisionally Induced Remote Charge Site Decompositions

Nancy J. Jensen[1], Kenneth B. Tomer[1], Michael L. Gross[1,3], and Philip A. Lyon[2]

[1]Midwest Center for Mass Spectrometry, Department of Chemistry, University of Nebraska, Lincoln, NE 68588
[2]Central Research Laboratories, 3M Company, St. Paul, MN 55144-1000

Several classes of FAB desorbed, closed-shell ions have been found to decompose upon collisional activation in an unprecedented manner. The fragmentation occurs for ions with long alkyl chains and involves parallel losses of the elements of C_nH_{2n+2} initiated from the alkyl terminus. The mechanism appears to be a 1,4-elimination of H_2 to give a neutral alkene (C_nH_{2n}) and an unsaturated charged fragment. The reactions do not rely on charge initiation and, as a result, they are termed "remote charge site fragmentations." Considerable structural information may be obtained by interpreting the spectra of daughter ions. The information includes location of double bonds in fatty acids, indentification of components in complex lipids, determination of compositions of anionic and cationic surfactants, and identification of long-chained alkyl substituents on phosphonium and ammonium ions.

Fast atom bombardment (FAB) and secondary ion mass spectrometry (SIMS) are important methods for structural determination of compounds. The methods are now being applied to more and more complex substances of biological and commercial interest. These compounds, when desorbed by using particle bombardment, may not give abundant fragmentation. Often, those fragment ions which are formed are difficult to distinguish from matrix ions when liquid matrices such as glycerol are used. Moreover, the compounds may occur as part of mixtures which may be difficult to separate prior to the mass spectrometry determination.

These problems can often be solved by employing a tandem mass spectrometer (MS-MS) as the mass analyzer for SIMS or FAB. The first mass spectrometer serves as a separations device which allows any given desorbed ion to be isolated for further study. Usually the ion is collisionally activated by accelerating it into a

[3]Author to whom correspondence should be directed.

0097-6156/85/0291-0194$06.00/0
© 1985 American Chemical Society

collision cell containing an inert gas, usually helium. One of the consequences of collisional activation is production of fragment ions which can be separated and detected by using the second mass spectrometer.

A few years ago, we began a research program to develop methods of analysis which would involve the use of FAB and a high performance tandem mass spectrometer. The tandem instrument was the first triple sector mass spectrometer to be designed and built by a commercial instrument company (Kratos of Manchester, U.K.). The first mass spectrometer of the combination is a double focussing Kratos MS-50 which is coupled to a low resolution electrostatic analyzer, which serves as the second mass spectrometer (1). This FAB MS-MS combination has been used to verify the structures of an unknown cyclic peptide (2), a new amino acid modified by diphtheria toxin (3), and an ornithine-containing lipid (4). A number of methods have also been worked out which rely on this instrumentation. They include the structural determination of cyclic peptides (5), nucleosides and nucleotides (6), and unsaturated fatty acids (7) and the analysis of mixtures of both anionic (8) and cationic surfactants (9).

In this chapter, we have chosen to review the results obtained in studies of the collisional activation of saturated and unsaturated fatty acid carboxylates and other related anionic substances such as sulfates and sulfonates. We will show that these kinds of compounds undergo a unique set of fragmentation reactions which occur remote from the charge site. Results which point to the mechanism of the reactions and to possible analytical applications, particularly the study of surfactants, will also be reviewed. These reactions are not reserved for negative ions only; certain positive ions also undergo remote charge site fragmentations. Our intention is to discuss in some detail one class of decomposition reactions in order to show the interesting chemistry which can be exhibited by gas-phase ions desorbed by particle bombardment. Moreover, we hope that the discussion will help stimulate new developments in desorption ionization which will permit investigations of this nature to be conducted on higher mass ions and lower quantities of chemical compounds.

Description of the Phenomenon

Fast Atom Bombardment (10) is an effective desorption method for many molecules of biological interest such as long chained fatty acids. In the negative ion mode, the FAB spectra of typical fatty acids show principally $(M-H)^-$ ions. The palmitic acid spectrum in Figure 1 is a representative example. Even under conditions of high multiplier gain and magnification, convincing evidence for additional fragmentation is lacking in spectra acquired into a computer at slow scan rates of 30 sec/decade.

While the spectrum shown in Figure 1 may be very useful in many applications, it does not reveal structural information. In many cases, such information may be obtained by studying the metastable decomposition of major ions. The instrument used in this laboratory may be operated in a manner in which an ion such as the $(M-H)^-$ ion of the fatty acid is selected, even at high resolution if necessary, by using MS-I and its decompositions in the field free region between MS-I and MS-II followed by scanning the second ESA (MS-II). However, the metastable decompositions of the $(M-H)^-$ ions of the fatty acids we have studied involve only a very low abundance water loss. This loss is only apparent for acids with chains of 16 carbons or more, and then the relative abundance of the appropriate metastable ions decomposing in the third field free region is a factor of 60,000 smaller than the abundance of the (M-H)⁻ ion.

However, high energy collisional activation (11,12) of carboxylate anions such as the (M-H)⁻ ion of palmitic acid, in an MS/MS experiment causes fragmentation as shown in Figure 2A. This distinctive and highly reproducible pattern is found for carboxylate anions with a carbon chain length of at least six or seven atoms and is well-defined for those anions containing 10 carbon atoms or more. The decompositions observed may be formally described as a series of parallel alkane losses from the (M-H)⁻ ion. Also observed are losses which lead to ions m/z 58 and m/z 86 and the loss of water. These decompositions occur for 1-2% of the collisionally suppressed ion beam. The series of parallel losses of the elements of neutral alkanes is a particularly unique feature. While the losses of neutral alkane fragments have been reported by several groups for a variety of compounds which include secondary alcohols, ketones, amines, and quaternary ammonium ions (13-19) these losses are associated with specific rearrangements at the charge site and are not part of a series of alkane losses. Another noteworthy feature of the CAD spectra is the abundance of high mass fragment ions. Stenhagen's classic studies of long chain carboxylate ester radical cations (20-22) show that fragmentations occur which favor the formation of low mass ions.

Mechanism. The abundance of high mass ions combined with the expectation that the charge should reside at the carboxylate site, suggest that the losses of the elements C_nH_{2n+2} preferentially occur from the end of the carbon chain remote from the carboxylate. This assumption was shown to be valid for the CH_4 loss from palmitic acid by comparing the CAD spectra of the (M-H)⁻ ion of unlabeled palmitic acid with the spectra of the (M-H)⁻ ion of 16,16,16-d₃-palmitic acid (Fig. 2). The two spectra differ only in the mass of the (M-H)⁻ ion. All other ions have the same m/z values indicating that indeed the CH_4 loss is from the alkyl terminus of the palmitate anion. Studies of other labeled carboxylic acids (23,24) including 7,7,8,8-d₄-palmitic acid, 9,10-d₂-myristic acid, 9,10-d₂-stearic acid, 2,3-d₂-octanoic, 9,10,12,13-d₄-stearic acid, 9,10,12,13,15,16-d₆-stearic acid, 4,5,7,8,10,11,13,14,16,17,19,20--d₁₂-docosahexanoic acid, and 5,6,7,8,11,12,14,15-d₈-eicosanoic acid, provide additional supporting evidence for the mechanism of this process.

The pattern of apparent alkane losses from specifically deuterated carboxylate ions is most easily rationalized by invoking a 1,2-elimination mechanism (see Equation 1)

$$CH_3(CH_2)_nCH_2\quad CH_2-(CH_2)_mCOO^- \longrightarrow CH_3(CH_2)_nCH_2CH_3 \qquad (1)$$
$$+ CH_2=CH(CH_2)_mCOO^-$$

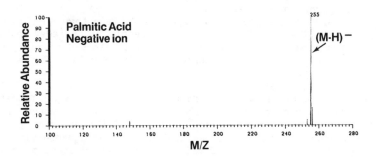

Figure 1. Negative ion mass spectrum of the FAB desorbed palmitic acid. Ion m/z 148 is from the triethanolamine matrix.

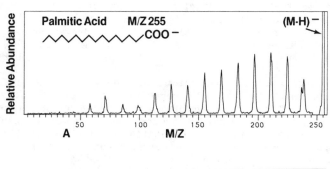

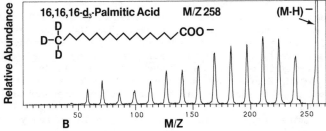

Figure 2. Spectra of the daughter ions produced by collisionally activating the (M-H)⁻ ions of unlabeled palmitic acid (A) and 16,16,16-d₃-palmitic acid (B). Note the spectra differ only in the mass of the main beam.

For example, upon collisional activation, 7,7,8,8-$\underline{d}_4$-palmitic acid loses unlabeled C_2H_{2n+2} from CH_4 to C_6H_{14}. However, the losses of seven and eight carbons are as $C_7H_{15}D$ and $C_8H_{17}D$. These losses of $C_7H_{15}D$ and $C_8H_{17}D$ are greater than 90%; $\underline{i.e.}$ less than 10% losses of C_7H_{16} or C_8H_{18} occur. This was most clearly established by using the NIEHS four sector mass spectrometer which has the capability for unit resolution of MS-II (we acknowledge Dr. Ron Hass for allowing us this opportunity).

    The 1,2-elimination mechanism as shown in Equation 1 is a four-electron orbital symmetry forbidden process ($\underline{25}$). Therefore, an alternative mechanism, such as the allowed six electron process shown in Equation 2, may be more likely. If the reaction procedes in this manner, a terminally unsaturated fatty acid carboxylate will be formed just as for the 1,2-elimination. The deuterium labeling

$$CH_3(CH_2)_nCH \quad \diagup\!\!\!\diagdown \quad CH\text{-}(CH_2)_mCOO^- \longrightarrow H_2 + CH_3(CH_2)_nCH_2=CH_2 \quad (2)$$
$$CH_2\text{-}CH_2 \qquad\qquad + CH_2=CH(CH_2)_mCOO^-$$

results will be the same for this mechanism as for that shown in Equation 1; that is, deuterium labeling will not permit us to distinguish between the mechanisms shown in Equations 1 and 2. The neutral products of the reaction will be hydrogen and an alkene, which, of course, cannot be distinguished mass spectrometrically from the alkane product required if the mechanism in Equation 1 is correct.

    In addition to mechanism considerations, several other generalizations can be made regarding the collisionally-induced dissociation of the long chained acids. The first is that this unique fragmentation behavior occurs for closed-shell carboxylate anions. The second is that the initiation of fragmentation remote from the charge site is a significant departure from the well-accepted charge-site or radical-site initiation of decompositions of gas-phase ions. The role of the charge site may be reduced due to the fact that this fragmentation appears to resemble a high energy thermal process. Evidence for this includes the fact that a series of parallel C_nH_{2n+2} losses occur in the pyrolysis of fatty acid esters. For example, Sun $\underline{et}$ $\underline{al}$. have shown that the pyrolysis of 9,10-$\underline{d}_2$ octadecanoate ($\underline{26}$) leads to production of a series of alkenes and unsaturated fatty esters. The relative abundances of the various unsaturated esters are comparable with the abundances of the product ions that we see in the CAD spectrum. The 1,4-elimination mechanism proposed above is entirely consistent with the results from the thermolysis studies.

    Other evidence that the fragmentation is a high energy process is found by obtaining mass spectra at a higher dynamic range than was employed for the full scan mass spectra discussed earlier. In the mass range m/z 224 to the molecular ion, ions m/z 225, 237, and 239 were seen with abundances of 0.64%, 0.29%, and 0.81% with respect to the (M-H)$^-$ ion at m/z 255. These ions correspond to the losses of C_2H_6, water and CH_4, respectively. Under normal FAB conditions, one expects the desorbed ions to have an energy distribution with at least a small number of ions of relatively high energy. As the narrow scan data indicate, a small fraction of the desorbed ions possess sufficient energy to undergo these remote charge site fragmentations. The phenomenon becomes much more apparent after high energy collisional activation.

<u>Evidence for Remote Site</u>. Losses of the elements C_nH_{2n+2} could occur remote from the charge site but still result from coiling (<u>27</u>) of the molecule to allow interaction between the alkyl chain and charge site. Evidence that coiling is not necessary for the reaction to occur was provided by examining the collisionally activated decomposition of the $(M-H)^-$ ion of cholesteryl hemisuccinate (Fig. 3), a rigid molecule. In this case C_nH_{2n+2} losses occur from the alkyl chain remote from the charge-bearing carboxylate. Analysis of a molecular model indicates that a minimum length of seven carbons in the alkyl chain is necessary for any interaction with the charge site. The lack of aromatic character or multiple double bonds in the steroid moiety are taken to discount the possibility of charge dispersion by resonance effects.

The CAD spectrum of the $(M-H)^-$ ion of cholesteryl hemisuccinate (Fig. 3) also illustrates one specific type of structural information which can be obtained from the remote loss pattern. The collisionally activated $(M-H)^-$ ions decompose by losing CH_4, a major fragmentation, and then the elements of C_3H_8, but virtually no loss of C_2H_6 occurs. This is readily explained in terms of the structure of the alkyl chain. Since there is a methyl branch at the end, increased probability for CH_4 loss and, according to the above mechanisms, reduced probability for C_2H_6 loss are expected. Hence, branch points in a carbon chain may be identified by the suppression of specific C_nH_{2n+2} losses. As indicated in this spectrum, remote losses resume once the branch point is passed; <u>i.e.</u> C_3H_8 and C_4H_{10} losses occur as expected.

<u>Strategy for Locating Double Bonds</u>. A structural feature revealed by the series of C_nH_{2n+2} losses from collisionally activated ions is the presence and location of a double bond in the alkyl chain of a carboxylate anion (<u>7</u>). This would be expected from either a 1,2-elimination mechanism proposed in Equation 1 or a 1,4-elimination of H_2 (Equation 2) as the transfer of vinylic hydrogens and cleavage of the double bond are not anticipated. The double bond is located by the absence of specific C_nH_{2n+2} losses. The collisionally activated decompositions of 9-hexadecenoic acid gives rise to a typical spectral pattern (Fig. 4). Two abundant ions (A and A′) are followed by three very low abundance ions at lower mass and then a third abundant ion (B). The abundant ions A and B arise from a hydrogen transfer and cleavage of the two C-C allylic bonds, one on the acid side and the other on the hydrocarbon side. Alternative and more likely mechanisms are given in Equations 3 and 4. The first mechanism depicts C_nH_{2n+2} loss remote from the double bond; the second shows the facile allyl-cleavage/rearrangement. Both should be thermally allowed.

$$CH_3(CH_2)_n CH \qquad CHCH=CH(CH_2)_m CO_2^- \longrightarrow \qquad (3)$$

$$CH_3(CH_2)_n CH=CH_2 + H_2 + CH_2=CHCH=CH(CH_2)_m CO_2^-$$

$$CH_3(CH_2)_n CH_2-CH \qquad CH-(CH_2)_m COO^- \longrightarrow \qquad (4)$$

$$CH_3(CH_2)_n CH_2 CH=CH_2 + CH_2=CHCH_2(CH_2)_m COO^-$$

The potential for locating multiple double bonds in long-chained carboxylate anions would be expected. However, the actual spectra obtained for such ions, are not interpreted easily as those of monounsaturated fatty acids. Furthermore, highly unsaturated acid anions undergo only loss of 45 amu, and no remote charge site fragmentation is seen. Labeling procedures, used initially in the mechanistic studies, provide an alternative approach for locating multiple double bonds. The labeled compounds, 9,10-$\underline{d_2}$-stearic acid, 9,10-$\underline{d_2}$-myristic acid and 2,3-$\underline{d_2}$-octanoic acid, were prepared by the reduction of their respective unsaturated compounds by using diimide (N_2D_2) (28,29). Collisional activation of the reduced compounds was not only informative regarding the mechanism but also permitted the position of the double bond to be determined by accounting for mass shifts of specific ions with respect to their unlabeled counterparts. The decreased intensity of peaks attributed to ions formed as a result of decompositions at labeled carbon sites is due to transfer of both hydrogen and deuterium as a given labeled carbon atom has one of each. Hence these less intense, slightly broadened peaks are actually unresolved doublets which also serve as markers for the double bond location. The diimide reaction was subsequently employed to reduce multiply unsaturated fatty acids including linoleic, linolenic, eleostearic, arachidonic, and docosahexaenoic, which contain two, three, three, four, and six double bonds respectively.

The reduction procedure involved reacting deutero diimide, which was generated from hydrazine-$\underline{d_4}$ in D_2O, with the unsaturated fatty acids (28,29). The acid was first neutralized with lithium hydroxide and the resulting lithium salt dried. The salt was then dissolved in D_2O along with N_2D_4, and the solution heated to slow reflux for 3-4 days until bubbling ceased. The reduced acid was then extracted in diethyl ether and investigated using FAB and CAD mass spectrometry.

Arachidonic acid, which is a representative example, was reduced to give approximately 30% $\underline{d_8}$-eicosanoic acid. The corresponding (M-H)$^-$ was sufficiently abundant that it could be

selected using MS-I of the tandem mass spectrometer and cleanly
collisionally activated. The CAD spectrum can be found in reference
23. Alkane losses involving the incorporation of one deuterium are
first found as the losses of C_4H_9D and $C_5H_{11}D$. According to the
mechanisms, the most remote double bond from the carboxyl end must
be at position 14. Similarly, losses of $C_7H_{13}D_3$ and $C_8H_{15}D_3$;
$C_{10}H_{17}D_5$ and $C_{11}H_{19}D_5$; and $C_{13}H_{21}D_7$ and $C_{14}H_{23}D_7$ are interpreted to
locate the other three double bonds at positions 11, 8, and 5,
respectively.

Application to Complex Lipids.   Structural information may be
obtained for complex lipids as a result of remote charge site
fragmentation (23). The negative ion FAB spectra, as well as the
collisional activation spectra of the major ions, have been obtained
for a number of phospholipids, including phosphatidylcholines,
phosphatidylserines,   phosphatidylinositols,   cardiolipid,
phosphatidylethanolamine and phosphatidylglycerol. In the case of
phosphatidylcholines, for example, the full scan FAB spectra show
that three high mass ions are formed which correspond to losses of
various portions of the choline. In addition, ions which may be
attributed to carboxylate anions from the fatty acid chains are also
observed.   For   α-phosphatidylcholine,   β-stearoyl-γ-oleoyl
(structure a), the principal ions of the FAB spectra are m/z 773,
728, 702, 283, and 281. The constituent fatty acid carboxylates are
the most abundant ions formed in the decomposition of the desorbed
lipid.   Collisional   activation   indicated   that   these   anions
originate via decompositions of the three high mass ions.   By
collisionally activating the m/z 281 ions and comparing the CAD
spectrum of the fragments of m/z 281 formed from the lipid with
those of the (M-H)⁻ of authentic samples of oleic and vaccenic acids
it could be confirmed that one of the fatty acid constituents was
oleic and not vaccenic acid, for example. As expected, the CAD

$$
\begin{array}{l}
\quad\quad\quad O \\
\quad\quad\quad \| \\
H_2C\text{-}O\text{-}C\text{-}(CH_2)_7CH{=}CH(CH_2)_7CH_3 \\
\quad| \\
\quad\quad\quad O \\
\quad\quad\quad \| \\
HC\text{-}O\text{-}C\text{-}(CH_2)_{14}\text{-}CH_3 \\
\quad| \\
\quad\quad\quad O \\
\quad\quad\quad \| \quad\quad\quad\quad + \\
H_2C\text{-}O\text{-}P\text{-}O\text{-}CH_2CH_2N(CH_3)_3 \\
\quad\quad\quad| \\
\quad\quad\quad O
\end{array}
$$

<u>a</u>

spectra of oleic acid and the m/z 281 ion from the lipid are
characterized by suppressed fragmentations corresponding to C_8H_{18},
C_9H_{20} and $C_{10}H_{22}$ losses. For vaccenic acid, fragmentations which
lead to expulsion of C_6H_{14}, C_7H_{16}, and C_8H_{18} are suppressed.

Application to Other Negative Ions. Remote charge site fragmentation and its utility for revealing structural information are not limited to carboxylate anions. Other classes of anions including alkyl sulfates, alkyl ether sulfates, alkyl sulfonates, and n-acylated amino acids can be desorbed as $(M-H)^-$ and collisionally activated to undergo C_nH_{2n+2} losses. All of these compounds possess the common features of long alkyl chains and stable, highly localized anionic sites.

Many of these substances have substantial commercial significance as surfactants. The wide range of compounds found in any given surfactant creates a challenging problem if one wishes to analyze it. The nature of the mixtures is often obscure as the raw materials are invariably mixtures of fatty acids, alcohols, or hydrocarbon precursors. Lyon et al. (8) have demonstrated that FAB combined with MS/MS can be used to deal with these mixtures.

The analysis of a mixture of alkylsulfonates is representative of the success of this approach (8). The CAD spectrum of a FAB-generated molecular anion of an alkylsulfate, such as hexadecyl sulfate (Figure 5), allows one to verify that the material is a sulfate (ions m/z 80 and 96 which are SO_3^- and SO_4^-, respectively), to obtain the length of the carbon chain by counting peak manifolds from m/z 96 to the molecular ion, and to verify that the alkyl group is normal (not branched). The peak manifolds correspond to ions formed by remote charge site C_nH_{2n+2} losses. The losses are comparable with those seen for carboxylates (Fig. 2). In addition, the basic pattern of relative abundances is similar; i.e. C_3H_8 loss most favored with a decrease for more complex C_nH_{2n+2}.

It appears for both carboxylates and sulfates that the remote charge site fragmentation is first seen when the product ion has a chain length of four atoms attached to the charge site. For fatty acids, the series begins at m/z 99 (see structure b), and for sulfates, it starts with m/z 137 (see structure c). Ions m/z 86 for carboxylates and m/z 124 for sulfates are analogs. Both are presumably 5-membered ring radical anions which contain the functional group (structures d and e). A difference between the sulfates and carboxylates is the loss of water which occurs only for the carboxylate anion. This decomposition is probably not a remote charge site reaction.

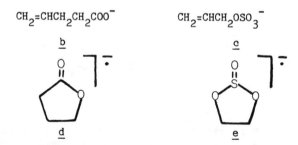

$$CH_2=CHCH_2CH_2COO^-$$ $$CH_2=CHCH_2OSO_3^-$$

b   c

d   e

As for the carboxylates, the remote site fragmentation is much more pronounced with increasing carbon chain length. For example, fragmentation is seen for octyl sulfate but is much more pronounced for longer-chained sulfates such as hexadecyl sulfate.

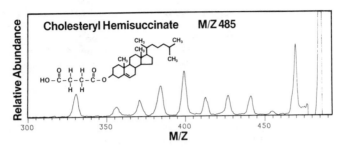

Figure 3. CAD spectrum of the negative ions from the dissociation of cholesteryl hemisuccinate (M-H)⁻ m/z 485. Expanded portion of the spectrum in the mass range m/z 300 to m/z 500 is shown.

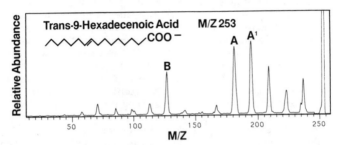

Figure 4. Spectrum of the daughter ions produced by collisionally activating the (M-H)⁻ ions of trans-9-hexadecenoic acid.

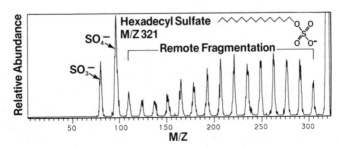

Figure 5. Spectrum of the daughter ions produced by collisionally activating the (M-H)⁻ ions of hexadecyl sulfate.

Nevertheless, the octyl sulfate CAD spectrum yields sufficient information to allow the compound to be distinguished from isomeric 2-ethylhexyl sulfate (for a comparison, see ref. 8). The branching of the 2-ethylhexyl sulfate causes suppression of fragmentation at the branch point in a fashion analogous to that seen for the carboxylates and for the steroid (see Figure 3).

Analysis of an alkyl ether sulfate surfactant further exemplifies the utility of the FAB MS/MS techniques for dealing with complex mixtures (8). The negative ion FAB spectrum contains a series of ions separated by 44 amu starting with m/z 265, the lowest mass ion of significant abundance, and continuing in a regular pattern to m/z 705. Additional homologous series differing by 14 amu are also observed. Selection and collisional activation of any of these ions produced consistent CAD spectra with two features: a C_nH_{2n+2} loss series and an equally well-defined repetitive pattern which could be attributed to each ethylene oxide unit (8). It was apparent from the CAD spectra and the relative abundances of the parent ions that the mixture included a series of dodecyl ether sulfates containing one to ten ethylene oxide units. Those with one to four units were found to be the most abundant. Other constituents containing tridecyl, tetradecyl, and pentadecyl alkane moieties gave ions which were superimposed on this dodecyl series. The remote C_nH_{2n+2} loss pattern of the respective CAD spectra allowed for identification of alkyl homologs.

The MS/MS approach to the analysis of surfactant mixtures seems to be applicable to most types of anionic surfactants (8). While each class exhibits certain unique fragmentations, which depend on the functional groups present, the characteristic negative ion CAD spectra reveal structural information pertaining to alkyl chain length and branching for all cases studied thus far except sulfosuccinates.

In cases of extensive branching, as for the alkylbenzenesulfonates, the C_nH_{2n+2} loss series is significantly perturbed compared to the straight chain alkyl sulfates, but it is sufficiently abundant to show perturbations in the loss pattern. These perturbations may be interpretable for identifying branch points. N-Acylated amino acids also show a suppressed remote charge site loss series because the preferred fragmentations are the decarboxylation of the parent anion and formation of the carboxylate anion of the amino acid (8).

Application to Ammonium and Phosphonium Ions. The remote charge site parallel C_nH_{2n+2} loss fragmentations are not limited to negative ions but have also been observed as a result of collisional activation of positive ions, including those formed from long chained amines, quaternary ammonium compounds, and phosphonium salts. These substances also share the common features of long alkyl chains and stable, closed-shell charge sites.

While remote site fragmentations are often the dominant decompositions of negative ions, positive ions usually undergo additional types of fragmentations. The CAD spectrum of hexadecylamine (Fig. 6) is a typical example of the spectra obtained for the (M+H)$^+$ ion of long chained amines. Ions at m/z 100 or greater result from fragmentations in which C_nH_{2n+2} segments are

expelled. This set of fragmentations is analogous to that seen for
the anions discussed previously in this chapter. Although they are
of low abundance, the ions may be used to determine chain length by
counting the number that appear in the CAD spectrum ($\underline{9}$).

As illustrated in Figure 6, two other types of fragmentation
are observed for collisionally-activated ammonium ions. The most
abundant products are alkyl and alkenyl cations which have three to
five carbon atoms. These presumably result from fragmentation of
higher mass aliphatic carbocations to yield the alkyl series of
C_nH_{2n+1} (m/z 43, 57, 71, etc.) and the alkenyl series of C_nH_{2n-1}
(m/z 41, 55, 59, etc.). For saturated, long-chained amines, the
predominant ions are of the alkyl series, but the presence of a
double bond causes the most abundant series to be alkenyl ions ($\underline{9}$).
The third fragmentation pathway yields nitrogen containing ions,
namely NH_4^+ (m/z 18) and $CH_2NH_2^+$ (m/z 30), and in some cases m/z 44.
This is in sharp contrast with the decompositions of amine radical
cations which lead principally to formation of nitrogen-containing
fragments.

The information from a full scan FAB spectrum and from CAD
spectra of selected ions may be used, as for the anions to analyze
a mixture and to obtain specific structural identifications. For
example, the positive ion FAB spectrum of dimethyldi-(hydrogenated
tallow)ammonium chloride has dominant ions at m/z 550, 552, 492,
466, and 438, which indicate that it is a mixture ($\underline{9}$). Collisional
activation of each of these ions, as shown for the m/z 494 ion in
Figure 7, reveals structural information. The increased prominence
of the remote site C_nH_{2n+2} loss fragmentation for dimethyl
quaternary ammonium ion as compared to amines (Fig. 6) is both
interesting and useful. For example, counting the number of
fragment ions (Fig. 7) from the molecular ion (m/z 494) to ions m/z
296 (A), m/z 268 (B), and m/z 240 (C), reveals that losses of
$C_{14}H_{30}$, $C_{16}H_{34}$ and $C_{18}H_{38}$, respectively, occur from the parent ion.
Since these ions terminate the prominent remote charge site loss
series, it is concluded that the mixture contains two dimethyl
dialkyl ammonium ions of the general formula $(CH_3)_2N^+R_1R_2$ where, for
one constituent, R_1 and R_2 are both hexadecyl and for the other, R_1
and R_2 are tetradecyl and octadecyl.

Mixtures of more complex nitrogen-containing compounds such as
amine oxides and ethoxylated quaternary amines are also amenable to
the type of investigation described above ($\underline{9}$). The CAD spectra of
these substances are dominated by a few informative ions which
result from specific fragmentations characteristic of the class of
compounds. The C_nH_{2n+2} loss series also occurs but at a less
abundant level for the ethoxylated compounds, and the series is of
low abundance for the amine oxides.

Another class of positive ions which undergo C_nH_{2n+2} losses
remote from the charge site are triphenyl alkyl phosphonium salts.
Collisional activation of n-decyl triphenyl phosphonium ions
desorbed directly by using FAB results in formation of daughter ions
which include a prominent ion m/z 262 attributed to $(C_6H_5)_3P^{\ddagger}$ and a
set of ions in the range m/z 300-387 formed by C_nH_{2n+2} losses
(Figure 8). A study of the dependence on translational energy
revealed that at high collision cell potentials (low collision
energies of a few hundred volts), remote site fragmentation is

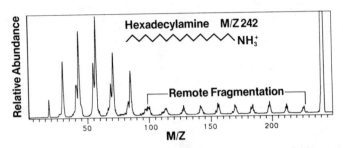

Figure 6. CAD spectrum of the decomposition of the $(M+H)^+$ ion of hexadecyl amine.

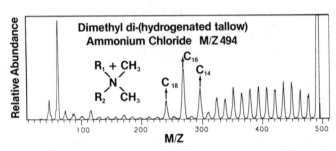

Figure 7. CAD spectrum of the positive ions from the dissociation of dimethyl di-(hydrogenated tallow)ammonium chloride m/z 494. This is a mixture of two dimethyldialkylammonium ions where R_1 and R_2 are C_{14} and C_{18} or both C_{16}.

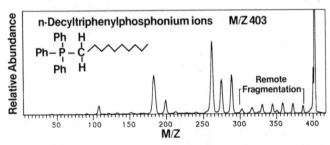

Figure 8. Spectrum of the daughter ions produced by collisionally activating positive n-decyltriphenylphosphonium ions m/z 403.

difficult to observe. However, it becomes an increasingly significant feature of the CAD spectrum as the collision energies are raised to kV levels (30).

Conclusion

Parallel losses of a series of fragments of the elements C_nH_{2n+2} from collisionally activated, closed-shell, long-alkyl chained ions is a wide-spread phenomenon. As shown, this type of fragmentation can be used to reveal structural features such as alkyl chain length and sites of branching and double bonds. Because the fragmentation requires high energy and does not rely on the charge site, we suggest that these decompositions constitute a new class of reactions not previously observed in mass spectrometry.

The fragmentation of cholesteryl hemisuccinate is a clear illustration that the phenomenon occurs physically remote from the charge site and that charge migration to the alkyl chain is apparently not important. Moreover, the ammonium ions, which have a covalently saturated charge sites, also show remote charge site fragmentation. The reactions of ammonium ions serve as further support for the idea that transitory charge on the alkyl chain, formed by hydrogen transfer to the charge site, does not initiate the fragmentation.

Improvements in SIMS and FAB which lead to more intense ion beams are highly desireable because that will permit the information to be obtained for smaller samples and for lower abundance constituents in complex mixtures such as surfactants. Moreover, consecutive activation steps (MS/MS/MS) should be important in investigations of mixtures of complex lipids and related materials. Here one step of collisional activation is necessary to liberate the fatty acid carboxylate and a second step is required to activate the anions. These experiments also require intense ion beams. It is our hope that the analytical possibilities raised by the chemistry discusses here will stimulate further research to improve FAB and SIMS.

Acknowledgments

The research reported here was supported by the National Science Foundation (Grant No. CHE 8320388), by the Midwest Center for Mass Spectrometry, an NSF regional instrumentation facility (Grant No. CHE 8211164), and by 3M Company.

Literature Cited

1. Gross, M.L.; Chess, E.K.; Lyon, P.A.; Crow, F.W.; Evans, S.; Tudge, H. *Int. J. Mass Spectrom. Ion Phys.* 1982, 42, 243-245.
2. Gross, M.L.; McCrery, D.; Crow, F.W.; Tomer, K.B.; Pope, M.R.; Ciufetti, L.M.; Knoche, H.W.; Daly, J.M.; and Dunkle, L.D. *Tet. Letts.* 1982, 23, 5381-5384.
3. Bodley, J.W.; Upham, R.; Crow, F.W.; Tomer, K.B.; Gross, M.L. *Arch. Biochem. Biophys.* 1984, 230, 590-593.
4. Tomer, K.B.; Crow, F.W.; Knoche, H.W.; Gross, M.L. *Biomed. Mass Spectrom.* 1979, 6, 356-358.
5. Tomer, K.B.; Crow, F.W.; Gross, M.L.; Kopple, K.D. *Anal. Chem.* 1984, 56, 880-886.

6. Crow, F.W.; Tomer, K.B.; Gross, M.L.; McCloskey, J.A.; Bergstrom, D.E. Anal. Biochem. 1984, 139, 243-262.

7. Tomer, K.B.; Crow, F.W.; Gross, M.L. J. Am. Chem. Soc. 1983, 105, 5487-5488.

8. Lyon, P.A.; Stebbings, W.L.; Crow, F.W.; Tomer, K.B.; Lippstreu, D.L.; and Gross, M.L. Anal. Chem. 1984, 56, 8.

9. Lyon, P.A.; Crow, F.W.; Tomer, K.B.; and Gross, M.L. Anal. Chem. 1984, 56, 2278-2284.

10. Barber, M.; Bordoli, R.S.; Sedgwick, R.D.; Tyler, A.M. J. Chem. Soc., Chem. Commun. 1981, 325.

11. Haddon, W.F.; McLafferty, F.W. J. Am. Chem. Soc. 1968, 90, 4745.

12. Levsen, K.; Schwarz, H. Angew. Chem., Int. Ed. Engl. 1976, 15, 509.

13. Veith, H.J. Mass Spectrom. Rev. 1983, 2, 419-445.

14. Hammerum, S.; Donchi, K.F.; Derrick, P.J. Int. J. Mass Spectrom. Ion Phys. 1983, 47, 347-350.

15. Denhez, J.P. Org. Mass Spectrom. 1983, 18, 131-132.

16. Gierlich, H.H.; Rollgen, F.W.; Borchers, F.; Levsen, K. Org. Mass Spectrom. 1977, 12, 388-390.

17. Sigsby, M.L.; Day, R.J.; Cooks, R.G. Org. Mass Spectrom. 1979, 14, 556-561.

18. Holmes, J.L.; Burgers, P.C.; Mollah, Y.A. Org. Mass Spectrom. 1982, 17, 127-130.

19. Marquestiau, A.; Meyrant, P.; Flammang, R. Bull. Soc. Chim. Belg. 1981, 90, 173-176.

20. Odham, G.; Stenhagen, E. Fatty Acids in "Biochemical Applications of Mass Spectrometry"; Waller, G.R. ed.; Wiley-Interscience: New York, 1972; pp. 211-227.

21. Dihn-Nguyen, N.; Ryhage, R.; Stallberg-Stenhagen, S.; Stenhagen, E. Ark. Kemi 1961, 393-399.

22. Ryhage, R.; Stenhagen, E. J. Lipid Res. 1960, 1, 361-390.

23. Gross, M.L.; Jensen, N.J.; Lippstreu-Fisher, D.L.; Tomer, K.B. "Proceedings of the International Symposium on Mass Spectrometry in Health & Life Sciences"; Elsevier, 1984 (in press).

24. Jensen, N.J.; Tomer, K.B.; Gross, M.L. J. Am. Chem. Soc., 1985 (in press).

25. Woodward, R.B.; Hoffman, R. Angew. Chem. Int. Ed. Engl. 1969, 8, 797.

26. Sun, K.K.; Hayes, H.W.; Holman, R.T. Org. Mass Spectrom. 1970, 3, 1035.

27. Meyerson, S.; Leitch, L.C. J. Am. Chem. Soc. 1971, 93, 2244-2247.

28. Thomas, A. "Deuterium Labelling in Organic Chemistry"; Appleton-Century-Crofts: New York, 1971; pp. 324-326.

29. Biemann, K. "Mass Spectrometry Organic Chemical Applications"; McGraw-Hill: New York, 1962, pp. 242-243.

30. McCrery, D.A.; Peake, D.A.; Gross, M.L. Anal. Chem., 1985 (in press).

RECEIVED April 16, 1985

Analysis of Reactions in Aqueous Solution Using Fast Atom Bombardment Mass Spectrometry

Richard M. Caprioli

University of Texas Medical School at Houston, Houston, TX 77030

Fast atom bombardment mass spectrometry has been utilized for the quantitative determination of ionic species, in glycerol/water solutions, which are produced by chemical and enzymic reactions. It is shown that reaction constants can be determined in this manner and that they can be accurately related to those determined by other methods used in the analysis of aqueous solutions. The reactions studied include proton dissociation constants for organic acids, an enzyme equilibrium constant, and enzyme rate constants using natural substrates.

Fast atom bombardment mass spectrometry (FABMS) has become an important addition to the ionization techniques available to the analytical chemist in recent years. It has been particularly useful in a number of diverse applications which include molecular weight determinations at high mass, peptide and oligosaccharide sequencing, structural analysis of organic compounds, determination of salts and metal complexes, and the analysis of ionic species in aqueous solutions. This paper will focus on some aspects of the quantitative measurement of ionic species in solution. The reader is referred to a more comprehensive review for more details of some of the examples given here as well as other applications (1).

One of the important questions with regard to the use of FABMS in following ionic reactions is whether the technique can accurately sample the ionic species in solution so as not to perturb the chemical dynamics which exist at that point in time, i.e., will the ions which are measured in the gas phase have the same ionic distribution as they had in the aqueous phase. If so, then under what conditions do these considerations hold?

A number of recent studies have shown that under certain conditions, FABMS indeed can very accurately measure the balance of ionic species in ongoing chemical reactions in solutions. These studies include the determination of acid dissociation constants (2), equilibrium constants for enzyme catalyzed reactions (1), metal-ligand association constants (3), and measurements of

0097-6156/85/0291-0209$06.00/0
© 1985 American Chemical Society

reaction rates for specific substrate–enzyme reactions ([4]).
Several of these applications will be discussed in this paper.

Acid Dissociation Constants

Early work with aqueous solutions containing ionic solutes in a 1:1
mixture of water and glycerol showed that factors such as the pH
of the solution and salt content had significant and reproducible
effects on the distribution of ionic species measured by the mass
spectrometer. Using the Henderson–Hasselbalch equation under
simplifying conditions (at low ionic strengths with acid components
whose pKa's lie between 3 and 10), it was shown that the pKa of an
acid could be accurately determined knowing the pH of the solution
and the concentrations of acid and base species ([2]). With respect
to the measurement of this constant by FABMS,

$$pKa = pH + \log \frac{(HA+H)^+}{(ANa+H)^+ + (ANa+Na)^+}$$

where $(HA+H)^+$ is the ion intensity of the undissociated acid HA,
and $(ANa+H)^+$ and $(ANa+Na)^-$ are the ion intensities of the corresponding conjugate base A^-. For example, the shift in the molecular
ion species as a function of pH is shown in Figure 1 for the
zwitterionic compound tris(hydroxymethyl)methylaminopropanesulfonate
(TAPS). As the pH is lowered below the pKa of the acid, the
protonated form of the compound predominates. Further, the quantitative shift of ion current can be described mathematically according
to established equations. Since these measurements were performed
in 50% glycerol solutions, one directly calculates the pKa' (G50)
for the acid, that is, an apparent pKa at a given ionic strength
in a solution containing 50% glycerol. In order to compare this
value with those reported in the literature, usually expressed as
pKa, it is necessary to apply corrections for the effect of ionic
strength and the dielectric constant difference between the aqueous
glycerol solution and water. The magnitude of these two effects
can be determined experimentally; a plot of the apparent pKa vs
ionic strength gives a straight line which can be extrapolated to
zero ionic strength and a plot of apparent pKa vs glycerol content
extrapolated to zero glycerol content. With these corrections, the
pKa's of several types of acids were measured and compared with
published values. For twenty-five different acids, the average
deviation of the value for the pKa determined by FABMS with
respect to those reported in the literature was approximately
±0.05 pKa units.

In further work involving the measurements of acidity constants
by FABMS, it was found that the measurements of pKa's could be
applied in a qualitative manner using the Born equation ([5]) to
calculate the average ionic radii of the acid and base species in
solution. A modified form of this equation follows ([6]),

$$(pKa)_s - (pKa)_w = \frac{122n}{} (1/D_s - 1/D_w)$$

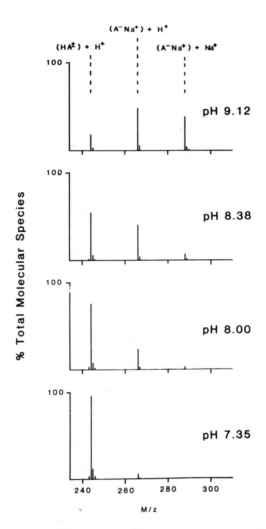

Figure 1. Effect of pH on the molecular ion species of solutions of TAPS [tris(hydroxymethyl)methylaminopropanesulfonic acid] in 50% glycerol/water. HA represents the protonated amine, a zwitterion of molecular weight 243, and A⁻ the conjugate base formed from dissociation of a proton from the acid. Reproduced from Ref. 2. Copyright 1983, American Chemical Society.

where the subscripts s and w refer to the values of the appropriate
constants in solvent and water, respectively, n is the charge number,
and r the average ionic radii of the ions. Essentially, the equation
shows that the change in the value of the pKa of a given acid in a
solvent relative to that in water is a function of the charge of the
various species, the radius of the ions, and the difference in the
dielectric constant of the two solvents. However, the equation is
not quantitative because it fails to take into account several
important factors, one of which is the energy of solvation of the
various ionic species. Nevertheless, the ionic radii calculated
from the pKa shift were found to approximate a straight line for
the mass range 50–500, with an acid of molecular weight 100 having
an ionic radius of approximately 2 angstroms and that of molecular
weight 400, approximately 7 angstroms. The importance of these
data lies not so much in the specific values measured, but rather
in the fact that they show that under specified conditions the
bombardment process does not substantially disrupt the balance of
ionic species so that the gas phase ionic distribution is similar
to that which existed in the liquid phase. Further, these distributions can be predicted from considerations derived from classical
approaches.

Several factors appear to be important for the accurate measurement of ionic components in aqueous solutions by FABMS. First, the
glycerol content can generally be varied between 40 and 70%; below
40% evaporation of the water from the FAB probe tip is too rapid to
obtain reproducible data and above 70% the dielectric constant change
is sufficiently large so as to significantly alter the acidity scale
and ionic interactions. Second, it is essential that sufficient
concentrations of counterions are present so that charge pairing can
occur, for example, all A^- ions should be paired to give $A^-..Na^+$.
If not, the base species will be inadequately collected in the gas
phase analysis by an instrument set up to collect positively
charged ions. Third, multiple positive charges on a molecule can
cause difficulty since these are generally more weakly charge paired
and lose a positively charged fragment either in the process of
sputtering from the sample surface or from fragmentation prior to
analysis.

Enzyme Reactions
<u>Enzyme Reactions</u>

Other types of reactions which have been studied using FABMS include
those catalyzed by enzymes. This application is particularly
interesting because it represents for the first time a generally
useful and molecularly specific probe with which to measure a
wide variety of enzyme substrates and products. Two approaches have
been successful, one in which the reaction is followed by the removal
of aliquots of sample taken at timed intervals with subsequent
analysis by FABMS and the other allowing the reaction to take place
in a glycerol-water mixture on the probe directly inside the mass
spectrometer. The choice of either method depends upon the
application. If the prime interest is to analyze a substrate, for
example, monitoring the release of amino acids from a polypeptide
using an exopeptidase, then direct analysis inside the spectrometer
may be preferred. If, on the other hand, the prime interest lies

in obtaining kinetic data for a particular enzyme-substrate reaction,
then analysis of a batch reaction where aliquots are removed at
timed intervals would be the better approach.

A number of enzymes have been shown to retain considerable
activity in aqueous-glycerol solutions and, further, can retain
this activity under intermittent bombardment with high energy
neutral atoms. These include trypsin, chymotrypsin, dipeptidyl
peptidases, proline specific endopeptidase, V8 protease, and
carboxypeptidase Y (4,7). Figure 2 shows the FAB mass spectrum of
the digestion products of the hydrolysis of the peptide β-casomorphin (Tyr-Pro-Phe-Pro-Gly-Pro-Ile) by the enzyme proline specific
endopeptidase. The ions at m/z 678 and 524 represent the (M+H)$^+$
species of the product peptides Tyr-Pro-Phe-Pro-Gly-Pro and Tyr-
Pro-Phe-Pro, respectively. This spectrum was taken after 17 minutes
of reaction within the mass spectrometer. Perhaps the most important aspect of this work is that it allows one to directly follow
in a real-time analysis the release of reaction products during
enzyme digestion of a substrate and to rapidly obtain structural
information even with very small sample sizes. In this regard, the
method provides an extremely sensitive, rapid, and mass specific
detection system.

Enzyme Kinetics

In the determination of steady state reaction kinetic constants
of enzyme-substrate reactions, FABMS also provides some very
unique capabilities. Since these studies are best performed in
the absence of glycerol in the reaction mixture, the preferred
method is that which analyzes aliquots which are removed from a
batch reaction at timed intervals. Quantitation of the reactants
and products of interest is essential. When using internal
standards, generally, the closer in mass the ion of interest is to
that of the internal standard, the better is the quantitative
accuracy. Using these techniques in the determination of kinetic
constants of trypsin with several peptide substrates, it was found
that these constants could be easily measured (8). FABMS was used
to follow the decrease in the reactant substrate and/or the
increase in the products with time and with varying concentrations
of substrate. Rates of reactions were calculated from these data
for each of the several substrate concentrations used and from
the Lineweaver-Burk plot, the values of Km and Vmax are obtained.
For example, Figure 3 shows the Lineweaver-Burk plot for the
hydrolysis of the peptide Met-Arg-Phe-Ala by trypsin. The X-inter-
cept is equal to -1/Km, and the Y-intercept to 1/Vmax. From these
data, the value of Km was found to be 1.9 mM, Vmax 0.31 μmoles/
ml/min, and kcat 7.1 sec^{-1} for this specific peptide.

The molecular specificity of FABMS opens new areas for
kinetic analysis of enzyme-substrate interactions. Since the
method is applicable to virtually all substrates whether or not
they have a UV or visible spectrum, natural substrates can be
used in place of synthetic substrates. Although much has been
learned through use of the latter, their reaction constants can
indeed be quite different than those of natural substrates.

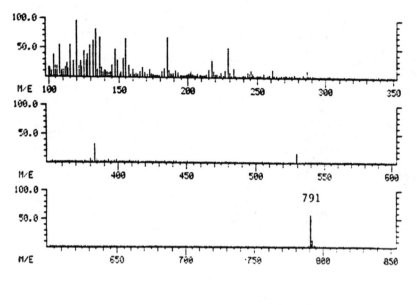

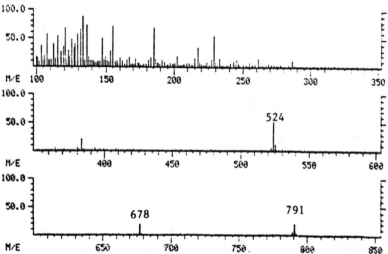

Figure 2. Hydrolysis of β-casomorphin, $(M+H)^+$ = 791, by proline specific endopeptidase in real-time within the mass spectrometer. Top: FAB spectrum before addition of enzyme. Bottom: FAB spectrum after 17 minutes of enzyme hydrolysis. Ions at m/z 678 and 524 correspond to new peptides produced in the reaction on the probe tip. Reproduced with permission from Ref. 8. Copyright 1983, L. A. Smith, Ph.D. Thesis, University of Texas Medical School.

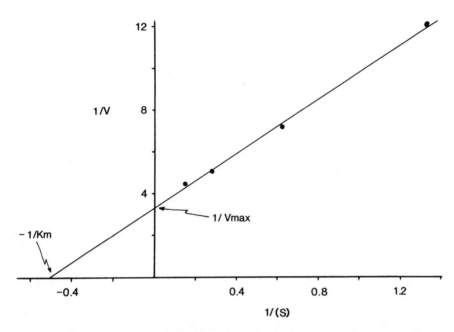

Figure 3. Lineweaver-Burk plot for the hydrolysis of Met-Arg-Phe-Ala by trypsin. The experimental parameters are described in reference 8. The velocity of the reaction, V, is expressed in units of μmoles/ml/min. and the substrate concentration, (S) in mM.

Conclusion

FABMS offers many advantages to the chemist in the analysis of
reactions taking place in aqueous solutions as well as for the
determination of ionic species under more static conditions. It
brings to bear on those applications where reactions must proceed
in aqueous solutions the molecular specificity of mass spectrometry.
The extraction, concentration and derivatization steps formerly
required for the analysis of organic compounds in aqueous solutions
using other mass spectrometric ionization methods are often difficult
and time consuming, but more importantly can lead to significant
changes in the chemical dynamics which exist in the aqueous media.
    One of the advantages of FAB over other soft ionization
techniques is that it uses glycerol (or some other suitable liquid
phase) which sets up a condition where the surface of the droplet is
constantly replenished with sample molecules. A second advantage
of the liquid matrix is that it allows solution chemistry to take
place virtually at the site of analysis.
    There is no doubt that FABMS has brought to the analytical
chemist an important tool with which to probe molecular structure.
It is a simple technique which can be employed on most mass spectrometers in use today. The near future promises utilization of the
method in the investigation of important new concepts in biochemical
and medical research.

Literature Cited

1. Caprioli, R. M. Specialist Periodical Reports 1984, 8.
2. Caprioli, R. M. Anal. Chem. 1984, 55, 2387.
3. Johnstone, R. A. W., Lewis, I. A. S.; Rose, M. E.
    Tetrahedron 1983, 39, 1597.
4. Smith, L. A.; Caprioli, R. M. Biomed. Mass Spectrom.
    1983, 11, 392.
5. Born, M. Z. Physik 1920, 1, 45.
6. Bates, R. G. Determination of pH, Theory and Practice,
    Wiley, New York, 1973; p. 211.
7. Caprioli, R. M.; Smith, L. A.; Beckner, C. F. Int. J. Mass
    Spectrom. Ion Phys. 1983, 46, 419.
8. Smith, L. A. Ph.D. Thesis, University of Texas Medical School
    at Houston, 1983.

RECEIVED April 16, 1985

Applications of Fast Atom Bombardment in Bioorganic Chemistry

Dudley H. Williams

University Chemical Laboratory, Lensfield Road, Cambridge CB2 1EW, England

The advent of FAB mass spectrometry has allowed the
routine molecular weight determination of polar
molecules, without derivatization, up to ca 3,000
Daltons, and in exceptional cases, within 1 mass unit
to the region of 8,000 Daltons. This advance, coupled
with FAB fragmentation, and enzymic digestion
techniques, has allowed the rapid solution of a number
of problems in protein and peptide chemistry - problems
which were hitherto rather difficult to solve. Examples
are given.

The discovery of FAB mass spectrometry in 1981 by Barber and co-
workers[1] has, along with SIMS,[2] and Californium plasma desorption
mass spectrometry,[3] revolutionized the study of polar molecules by
mass spectrometry. This paper is concerned with the application of
FAB mass spectrometry to solve problems in peptide and protein
chemistry.

The work to be presented has, when done in the author's
laboratory, been carried out on a Kratos MS-50 mass spectrometer,
fitted with a magnet possessing a mass range of 3000 Daltons at
full accelerating voltage (8KV). Xenon atoms of 4-9 KeV
translational energy have been used as bombarding particles into
a matrix most commonly consisting of thioglycerol/diglycerol (1:1),
or glycerol. The instrument is fitted with a post-acceleration
detector, so that the translational energies of incoming ions can

0097–6156/85/0291–0217$06.00/0

be increased by 9KeV, before impinging on an aluminized button. This device is indicated schematically in Fig. 1.

The most significant advantages of the FAB method are the ability to study underivatized peptides, a reduction in sample size, and the facility to study larger peptides. Peptides containing between 4 and 30 amino acid residues are conveniently studied with our present equipment.

A sample size of approximately 0.1 nanomoles is generally sufficient for molecular weight determination in either the positive or negative ion mode, but does not normally allow the sequence of amino acids to be determined. Larger sample sizes, typically between 1 and 5 nanomoles, afford some sequence information. Sequence ions are observed in the positive and negative ion modes from both N- and C-termini of the peptide and this may enable the complete sequence of the peptide to be determined.

The amount of sequence information available in the positive and negative ion FAB mass spectra of peptides varies considerably. The abundance of sequence ions is in part dependent on sample size but this is clearly not the only factor since large sample sizes (50 nanomoles) often fail to provide extensive sequence information for some peptides. Sequence information is frequently not complete and the absence of either C or N terminal sequence ions does not allow the complete sequence determination.

EI mass spectrometry of derivatized peptides complements FAB mass spectrometry but suffers from the disadvantage that larger sample sizes are usually required. For sample sizes less than 20 nanomoles and for high molecular weight peptides, alternative methods of generating sequence information are desirable. Carboxypeptidase digestion to generate a mixture of C-terminal fragments and chemical removal of N-terminal amino acid residues via the subtractive Edman degradation, coupled with FAB mass spectrometry offer a promising alternative.[4]

As an example where sequence information is available without the use of Edman or carboxypeptidase degradation, a study of calcineurin B may be cited. This work also allowed, most importantly, the determination of an N-terminal blocking group as myristic acid.

Calcineurin B is a calcium-binding protein first found in

bovine brain, and early work on this protein demonstrated that the NH$_2$-terminus of calcineurin B was blocked. When calcineurin B was cleaved with cyanogen bromide, and applied to an HPLC column, a peptide (CB-1) lacking a free NH$_2$-terminal amino acid was eluted as a sharp peak at 57% acetonitrile. The amino acid composition of CB-1 showed that it was a decapeptide containing only two hydrophobic amino acids. This indicated that the blocking group was much more hydrophobic than the acetyl, formyl or pyrrolidone carboxylic acid groups most commonly found at the N-termini of proteins.

Peptide CB-1 had M_r 1271 as determined by FAB mass spectrometry in the positive ion mode [$(M + Na)^+$ = 1294, $(M + H)^+$ = 1272] and in the negative ion mode [$(M - H)^-$ = 1270]. The number of carboxylic acid groups was determined from the positive ion FAB mass spectrum of the esterified peptide. An increase in M_r of 60 was observed, which corresponds to the formation of two methyl esters plus methanolysis of the C-terminal homoserine lactone. The M_r of the amino acid component of CB-1 was 1060, indicating that the M_r of the blocking group was 211.

When calcineurin B was digested with Staphylococcus aureus proteinase, a peptide, SP1, lacking a free NH$_2$-terminal amino acid was also eluted from HPLC at 57% acetonitrile. Amino acid analysis showed it to be a tripeptide Gly,Asx,Glx. From the known specificity of S. aureus proteinase the C-terminal reside of SP1 must be Glu. FAB mass spectrometry established the M_r of SP1 as 528, and esterification of this peptide led to an increase in M_r to 556, which corresponds to the formation of two methyl esters. Since SP1 has two free carboxyl groups, the sequence of SP1 must be X-(Gly,Asn)-Glu and the M_r of the blocking group must be 211.

The assignments of sequence ions observed in the FAB mass spectrum of CB-1 and its methyl ester are given in Table I, and suggested that the sequence of CB-1 was:

Gly-Asn-Glu-Ala-Ser-Tyr-Pro-Leu-Glu-Hsl

Assuming that the blocking group X is linked to the NH$_2$-terminal glycine by the usual amide bond, then hydrolysis should yield a carboxylic acid M_r = 228. This corresponds to the M_r of a C_{14} saturated fatty acid. the mass spectra are consistent with the supposition that X is $C_{13}H_{27}CO-$. This possibility was confirmed by

Table I A

Assignment of the sequence ions in the
FAB mass sectra of decapeptide

Type of fragment ion[a]	M_r-Values of fragment ions											M_r determining peak
	X	Gly	Asn	Glu	Ala	Ser	Tyr	Pro[b]	Leu	Glu	Hsl	
+ Na						621	708		968	1081	1210	1294
+ NAcy								856	953	1066		1294
+ CAmin	1027	913	784	713								1294
+ CAmin	1005	891	762									1272
+ CAlk		897	768									1294
− NA						597	684	b	944	1057	1186	1270
− NAcy						582		832	929	1042	1171	1270
− CAmin	1003	889	760	689								1270
− CAlk	988	874	744									1270

Most of the sequence ions observed in the positive ion (+) mode are cationised by NA⁺. The base peaks in the positive ion FAB mass spectra of peptides that lack basic functional groups as in CB-1 often correspond to the M_r of peptides cationised by the formation of adducts with Na⁺ and/or K⁺, traces of which are usually present in the matrix.

Table I B

Assignment of the sequence ions in the
FAB mass spectra of the decapeptide ester

Type of fragment ion[a]	M_r-Values of the fragment ions									M_r determining peak
	X	Gly	Asn	Glu-OMe	Ala,Ser	Pro[b]	Leu	Glu-OMe	Homo-Ser-OMe	
+ Na							982	1095	1238	1354
+ NAcy						870	1080			1354
+ CAmin	1087	973								1354
+ CAlk		957	814							1354

[a]Fragment ions are named as follows:

```
        CAmin                    CAlk
         ┌ H ↗                    ┌                 O        H ┐
─────────┤                ────────┤ CH ──── C ──────┤  N ──────┤
         └ N                      └ R               ‖          └
                                                    O
                                              NAcy       NA
```

Bond cleavages are accompanied by a hydrogen transfer to the charged fragment except in the NAcy case. Positive and negative signs indicate cationic and anionic fragment ions, respectively.
[b]Due to cyclic nature of the proline residue fragmentation of NA type does not produce fragment ions.

isolating the blocking group as its methyl ester. On gas chromatography, this ester co-migrated with methyl myristate.

We were also able to use FAB mass spectrometry to determine the
amino acid sequence around the active site serine in the acyl
transference domain of rabbit mammary fatty acid synthase.[6] The
synthase was labelled in the acyl transferase domain(s) by the
formation of O-ester intermediates after incubation with $[^{14}C]$-
acetyl- or malonyl-CoA (Fig. 2A). The modified protein was then
digested with elastase (Fig. 2B), and radioactive material isolated
via successive purification steps on Sephadex G-50 and reverse phase
HPLC. The isolated peptides were then sequenced by FAB MS. The data
summarized in Table II established the sequences of both the acetyl
and malonyl hexapeptides to be N-acyl-Ser-leu-Gly-Glu-Val-Ala.

The N-terminal location of the acyl group was confirmed by
showing that the molecular ions were identical after treatment with
acetic anhydride, which would acetylate a free amino group. Since
the linkages between the acetyl and malonyl groups and intact fatty
acid synthase are sensitive to hydroxylamine (1 M, pH 9.5, 2.0h,
$38^{\circ}C$; unpublished work), the initial acylation cannot be at an amino
group. An O→N migration must have occurred after proteolytic
digestion.

Our own work is now evolving, in part, to use FAB mass
spectrometry to determine which portions of some genes are expressed
in the frog Xenopus laevis. The genes in question are those coding
for some of the peptides which are secreted through the frog skin
when the frog is stimulated (by handling, or by infection with
epinephrine or nor-epinephrine). These peptides are significant
because of the remarkable structural, and physiological activity,
similarities which exist between them and a number of mammalian
neuropeptides. If known (and sequenced) frog skin peptides are used
as a guide to synthesize appropriate c-DNA probes, the genes which
code for these peptides can be isolated and sequenced. However, it
is then a problem to know which other parts of the gene are expressed in peptide production. FAB mass spectrometry has proved to be a
very powerful tool in solving this problem. FAB MS has been used
to show that a large number of hitherto unidentified peptides are
excreted by Xenopus laevis, although for most of these only MH^+ ions

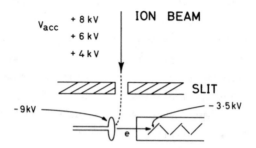

Figure 1. Schematic illustration of the post-acceleration detector.

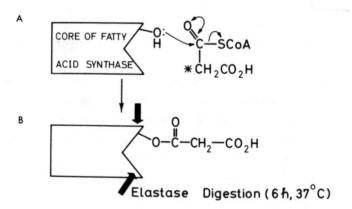

Figure 2. Schematic illustration of protein radio-labelling and elastase disgestion.

Table II

Amino acid sequence information[a] derived from negative ion FAB mass
spectra of the malonyl- and acetyl-hexapepties[b]

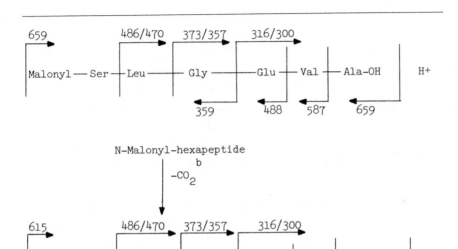

N-Malonyl-hexapeptide

N-Acetyl-hexapeptide

[a]Sequence ions observed in the negative ion FAB mass spectra of the
above peptides are as follows:

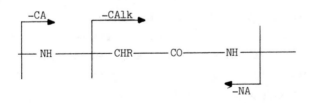

where the -CA and -NA cleavages occur with hydrogen transfer to the
charged fragment.
[b]Some decarboxylation of the malonyl group to produce an acetyl
group was observed in the FAB mass spectra of the N-malonyl-
hexapeptide.

are observed in peptide mixtures. However, when these molecular ions
are considered in the light of known gene sequences, and in the
light of possible processing sites in the peptides which they would
produce if transcribed and translated, full peptide sequences
corresponding to the molecular ions can be proposed. These peptide
sequences can then be validated (or otherwise) by one or two cycles
of Edman degradation, with re-determination of the molecular weight
by FAB (as outlined previously in this article) after each cycle.[7]
Although the above work is not completed, the principle by which FAB
mass spectrometry can be used in conjunction with known gene
sequences, and in particular to check their accuracy, can be
illustrated by reference to the recent work of Gibson and Biemann.[8]

It is obvious that it is prudent to check the correctness of
the amino acid sequence derived from the base sequence of the gene
not only at the NH_2 and COOH termini, which is the common practice,
but throughout the entire protein. This would help to uncover any
significant errors as well as address the possibility of post-
translational modifications.

For checking the correctness of a proposed structure the protein is hydrolyzed at specific sites - i.e., with trypsin - to
produce a pool of smaller peptides, which is then partially
separated by HPLC into five or six fractions. Each fraction is then
subjected to FAB-MS, which allows the determination of the molecular
weights of most or all of the peptides present in each fraction.
These values are then compared with the molecular weights of the
tryptic peptides predicted from the DNA deduced amino acid sequence.
The fractions can then also be subjected to Edman degradation(s) and
the molecular weights of the shortened peptides determined by FAB-MS
to identify the NH_2-terminal amino acid(s) of each peptide from the
change in its molecular weight. For example, a tryptic digest of the
protein Gln-tRNA synthetase was separated into six crude fractions
by HPLC to remove the enzyme, reagents, salts, and other contami-
nants and also to reduce the complexity of the original digest,
which, in turn, reduces the complexity of the resulting FAB mass
spectra. There were two predicted tryptic peptides of M_rs 736 and
1379, respectively, which is 10 daltons less than two unmatched experimentally found peptides (M_rs 746 and 1389). The former cover the

amino acid positions 7-12 and 1-12, i.e., they are overlapping peptides from the NH_2 terminus of Gln-tRNA synthetase. Only the pair serine/proline differs by 10 daltons. The preliminary sequence indeed contained serine in position 7, which is common to both, and the condon used for serine was TCG, which could be converted to proline condon CCG if base 19 were C rather than T.

Re-inspection of the DNA sequence analysis data revealed that not only base 19 is C but also base 18. Because both CGT and CGC code for arginine, the corrected sequence now contains an arginine-proline bond, which trypsin splits very slowly. This is probably the reason why both tryptic peptides (positions 1-12 and 7-12) were observed. The following are the preliminary and final DNA and amino acid sequences of this region.[8]

AGT-GAG-GCA-GAA-GCC-CGT-TCG-ACT-AAC-TTT-ATC-CGT
Ser-Glu-Ala-Glu-Ala-Arg-Ser-Thr-Asn-Phe-Ile-Arg
AGT-GAG-GCA-GAA-GCC-CGC-CCG-ACT-AAC-TTT-ATC-CGT
Ser-Glu-Ala-Glu-Ala-Arg-Pro-Thr-Asn-Phe-Ile-Arg

Conclusion

It is evident from the foregoing examples that FAB mass spectrometry is a powerful technique in several aspects of protein and peptide chemistry. This is particularly true in determining the nature of blocking groups in proteins, in checking the sequences of bases in genes (by examining the resulting protein), and in determining which portions of genes are expressed (when the peptide products are readily available as relatively pure components, or simple mixtures).

Acknowledgments

The author thanks the SERV (UK) for support of that portion of the work carried out at the University of Cambridge.

Literature Cited

1. M. Barber, R. S. Bondoli, R. D. Sedgwick, and A. N. Tyler, J. Chem. Soc. Chem. Commun., 1981, 325.

2. See, for example, H. Grad, N. Winograd, and R. G. Cooks, J. Am. Chem. Soc., 1977, 99, 7725.

3. R. D. MacFarlane and D. F. Torgerson, Science, 1976, 191, 920.

4. C. V. Bradley, D. H. Williams, and M. R. Hanley, Biochem. Biophys. Res. Commun., 1982, 104, 1223; see also Y. Shimonishi, Y. M. Hong, T. Kitagishi, T. Matsuo, H. Matsuda, and I. Katakuse, Eur. J. Biochem., 1980, 112, 251.

5. A. Aitken, P. Cohen, S. Santikarn, D. H. Williams, A. G. Calder, A. Smith, and C. B. Klee, FEBS Letters, 1982, 150, 314.

6. A. D. McCarthy, A. Aitken, D. G. Hardie, S. Santikarn, and D. H. Williams, FEBS Letters, 1983, 160, 296.

7. B. W. Gibson, L. Poulter, and D. H. Williams, unpublished work.

8. B. W. Gibson and K. Biemann, Proc. Natl. Acad. Sci U.S.A., 1984, 81, 1956.

RECEIVED April 24, 1985

Use of Secondary Ion Mass Spectrometry to Study Surface Chemistry of Adhesive Bonding Materials

W. L. Baun

Mechanics and Surface Interactions Branch, Air Force Wright Aeronautical Laboratories, AFWAL/MLBM, Wright-Patterson Air Force Base, OH 45433

Secondary Ion Mass Spectrometry used as a solo instrument or in concert with other methods has proven to be an excellent technique for studying the surface chemistry of adhesive bonding materials. The application of SIMS is shown in relation to pretreatments of metals and alloys, chemistry and structure of adhesives, and locus of failure of debonded specimens.

The idea of building structures which are stronger and more durable while at the same time lighter in weight would appear contradictive, but has been accomplished using adhesive bonding and composite materials. Such novel construction and materials are used extensively in the aerospace and automotive industries. Since these structures depend on the interaction of surfaces and the formation of interfaces, it is necessary to develop methods of physical and chemical characterization which are applicable to such types of materials. Secondary ion mass spectrometry, used either as a stand alone instrument, or as a complement to other techniques, has proven of value for characterization of original materials and failure surfaces following use or accelerated test. Since these proceedings contain detailed descriptions of the SIMS technique and its variations, emphasis will be placed in this account on application of the method to surface preparation and adhesive bonding. Theoretical and practical operational aspects of SIMS will be considered only insofar as they pertain to adhesive bonding research.

Discussion

In deciding which surface chemistry tools to use for the broad area of adhesion and for adhesive bonding in particular, a number of aspects must be considered. More often than not, a combination of instruments must be used to take advantage of the unique information provided by each method. Table 1 shows some of the important aspects of adhesive bonding and some of the characterization methods which are applicable in these areas[1]. The acronyms are those used in the review by Powell[2].

It is seen in Table 1 that SIMS is applicable to several areas
of investigation in adhesive bonding. SIMS may be used in a variety
of ways including species imaging of the surface (SIIMS) which may
be especially useful for clarifying mixed mode failure surfaces.
The main features of SIMS are shown in Table II[3].

Chemistry of Adherends

A determination of the chemistry of metallic adherends presents
problems of each of the areas discussed here. Many of the surface
chemical techniques are applicable to the analysis of adherends, and
because of the stability and good conductivity, decomposition and
surface charging are not problems. Surface chemical analysis is
usually devoted to (1) determining the amount and distribution of
elements purposely placed on the surface to impart a desirable
property, and (2) detection and monitoring of impurity elements
which may be deleterious to the adhesive bond. Many chemical
etching and oxidizing treatments are used on metal and alloys to
enhance adhesive bonding of the surface. Enhancement comes about by
roughening of the surface and by changing the surface chemistry. In
addition, some thermal treatments, such as the bond cure in adhesive
bonding, may affect the composition of the surface, either by
introducing impurities or by increasing or decreasing a concentration of alloying elements at the surface. McDevitt and co-workers[4, 5, 6] used SIMS and other modern surface analysis tools to
analyze several aluminum alloys following chemical treatment for
adhesive bonding. They found a number of interesting phenomena,
including the formation of an interfacial region rich in copper on
the structural alloy 2024 aluminum. The concentration and width of
this potential weak boundary layer was found to vary depending on
the etching conditions of the sulfuric acid-sodium dichromate
solution. This solution is related to the surface preparation
method known as the FPL etch. Similar results were obtained more
recently by Sun and Co-workers[7]. The formation of such potential
weak boundary layers may influence both the initial bondability and
the long time durability of the adhesive bond. Baun et al used Ion
Scattering Spectrometry (ISS), Secondary Ion Mass Spectrometry
(SIMS) and Auger Electron Spectrometry (AES) to analyze a variety of
metal and alloy adherends. These authors also used several surface
treatments on titanium and titanium alloys and analyzed them by
surface techniques such as ISS, SIMS, AES and SEM[9, 10, 11]. Large
differences in chemistry were observed on titanium and its alloys
depending on the surface treatments. Some possible steps in the
surface preparation of titanium alloys for adhesive bonding are
shown in Table III[12].

Table I
Aspects of adhesive bonding and applicable surface characterization methods[1]

Adherend chemistry
AEAPS, AEM, AES, APS, BIS, CIS, CL, EM, ES, EXAFS, IIRS, IIXS, IMMA, IS, ISS, LMP, PES, RBS, SIMS, SXAPS, SXES

Adherend structure and morphology
AEM, ELL, EM, HEED, IMMA, LEED, SEM, SIIMS, SRS, STEM, TEM, XEM, XRD

Adhesive chemistry
AES, AIM, ASW, ATR, ESR, HA, IRS, ISS, LS, PES, SIMS, UPS, XPS

Adhesive structure and morphology
ATR, IR, UV, RAMAN, SEM

Interaction of polymers with metals
AES, AIM, ASW, CPD, ELL, EELS, ESDI, ESDN, FD, FDS, HA, IRS, IR, ISS, ISD, LEED, LS, PD, SC, SIMS, UPS, XPS, RAMAN

Failure surfaces (locus of failure)
AES, ATR, ELL, ISS, SIMS, PES, XPS, SEM, SXES, SXAPS, SRS, UPS

Table II
Main features of SIMS as a surface analysis method

Positive	– Information depth in the "monolayer range"
	– Detection of all elements including hydrogen
	– Detection of chemical compounds
	– Isotope separation
	– Extremely high sensitivity for many elements and compounds ($\sim 10^{-6}$ monolayers)
	– Quantitative analysis after calibration
	– Negligible destruction of the surface (Static SIMS)
	– Elemental profiling (Dynamic SIMS)
	– Elemental and Cluster Imaging
Negative	– Large differences in sensitivity for different "surface structures" (factor 1000)
	– Problems in quantitative interpretation of molecular spectra
	– Ion induced surface reactions
	– Surface Charging in Insulators

Table III
Surface Preparations of Titanium Alloys for Adhesive Bonding[12]

1 Clean	2 Etch	3 Convert	4 Modify
Solvent	Acids	Chemical	Boiling H_2O
1. Liquid	HF, HNO_3,	Phosphates,	Boiling H_{2O} + ?
2. Vapor	H_2SO_4,	Fluorides,	Dry Heat
Abrasion	H_3PO_4	etc.	Heat + Humidity
Alkaline	Combinations	Anodization	Absorption
Combinations	Alkaline	dense	UV, Ions,
	Abrasive	porous	
	Slurry	Combinations	Corona, etc.

The surface chemistry of Titanium alloys varies with each physical or chemical treatment. An example of ISS/SIMS results from a typical chemical pretreatment for Ti-6Al-4V is seen in Figure 1. Here the sample was degreased, etched in HNO_3/HF and converted with a mixture of HF, NaF and Na_3PO_4 in aqueous solution. Spectra from this surface shows that it is far from being a simple oxide. An ISS spectrum from a typical TiO_2 surface is shown in the inset for comparison. Note in the SIMS data the appearance of the molecular ion TiF^+, which suggests the combination of fluorine and titanium. It is also interesting that sodium and fluorine do not seem to be associated even though appreciable amounts of each occur on the surface. SIMS data in Figure 2A for a typical anodized oxide formed in a neutral Na_2HPO_4 + H_3PO_4 solution at 50 volts shows a spectrum similar to the crystalline oxide, rutile. These thin anodized specimens show large amounts of TiO^+ in relation to Ti^+. It is interesting that the alloying element vanadium is nearly absent in the oxide layer. The surface chemistry of porous anodized oxides are much different as seen in Figure 2B where the SIMS spectrum is shown for a thick oxide formed in the same electrolyte as A, but at 100V which is near the breakdown voltage[13] for this electrolyte and titanium. SIMS data show hydrocarbons, alkali elements, calcium, and most importantly evidence of phosphorus (probably phosphate ion) in the porous film. AES elemental profiles show that phosphorus concentrations are high at the surface and continue on into the film.

Initial bondability of anodized surfaces was tested in the lap shear configuration using numerous commercial epoxy adhesives. With one exception, all surfaces proved to be bondable and gave accept-able lap shear values. That exception was an anodized film formed

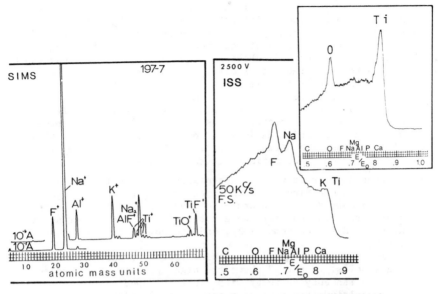

Figure 1.   ISS/SIMS Data for Ti-6Al-4V Alloy Etched with HF/HNO$_3$ and Converted with HF/NaF/Na$_3$PO$_4$. Inset shows typical ISS Spectrum from TiO$_2$.

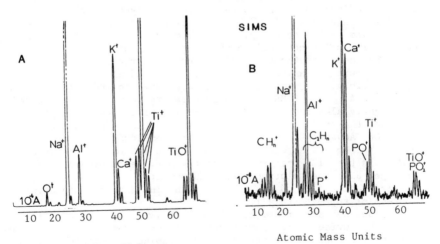

Figure 2.   SIMS Data for Ti-6Al-4V Alloy.
    A.   Specimen Anodized in Na$_2$HPO$_4$ + H$_3$PO$_4$ (pH = 7) at 50 Volts.
    B.   Specimen Anodized in Na$_2$HPO$_4$ + H$_3$PO$_4$ (pH = 7) at 100 Volts (Breakdown)

in any electrolyte containing fluorine ions. The result was rather
unexpected since a commercial patent calls for the addition of
fluorine ions to a solution to increase the current density and
subsequently the porosity of the anodized film for adhesive bonding.
The adhesive showed good adhesion to the oxide film in these cases.
Failure often occurred interfacially at the oxide/metal interface
even using lap shear specimens. When a test in which pure shear was
placed at the interface, as in the three-point-bend method, then
failure always took place interfacially at the oxide/metal interface. Peel tests on anodized films in which fluorine was present
also showed interfacial failure in most tests. Peel strength in the
anodized regions was virtually zero. SIMS spectra showed fluorine
to be present on these surfaces (both on the adhesive and adherend).

An interesting result was the appearance of F^+ in the residual gas
analysis when the electron beam in AES was placed on the sample,
suggesting easy desorption and a very unstable surface. In fact,
when electron beam currents were not minimized, the fluorine frequently was desorbed completely, and did not appear in the Auger
spectrum.

SIMS spectra from simple etching processes also showed interesting results. Figure 3 shows the high activity of etched surfaces, and tendency to react with elements found in tap water. Note
that the calcium from the water appears to combine with fluorine
left on the surface by the hydrofluosilicic acid, but there is
little suggestion of any reaction between fluorine and titanium to
form a compound. Also of interest is the very low concentration of
the alloying element aluminum on these surfaces in view of the high
secondary ion yield from aluminum. Vanadium, which was not observed
in the anodized specimens, appears prominently on most acid etched
surfaces. Aluminum alloys show equally interesting surface chemistry changes with processing. Many aluminum alloys following processing including hot rolling and heat treatment, show surface
elemental concentrations far different from true bulk composition.
Even following cleaning such as degreasing and alkaline bath,
appreciable differences are seen between surface and bulk as shown
in Figure 4. Here SIMS and ISS spectra are shown for a degreased
and lightly alkaline cleaned 2024 alloy. SIMS shows a large amount
of Mg on the surface and the ISS ratio of 0 to Mg-Al is about that
MgO. One advantage of SIMS showing its complementary nature is seen
here where Mg and Al cannot be resolved in ISS but is easily separated in the SIMS spectrum. When the surface is etched in a
stronger alkaline solution, the SIMS spectrum (B) shows a much
smaller ratio of Mg to Al, much more in line with the magnesium
content of approximately 1.5%.

Other surface treatments which etch away the surface still
leave the surface composition much different from the bulk. Figure
5 shows the ISS/SIMS spectra for 2024 aluminum alloy etched in a
mixture of nitric and hydrofluoric acids. As is seen in both
spectra, copper is prominent on the surface. This is a very mild
case of surface smutting[8]. Surface smut is observed in many materials which are heavily etched in acid or alkaline media. Smut on
stainless steel has been studied by ISS/SIMS[14]. An example of such
spectra on a smutted 304 stainless steel surface is seen in Figure 6

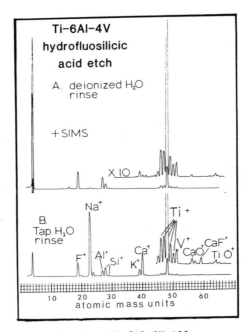

Figure 3.   SIMS Data from Ti-6Al-4V Alloy.
A. Etched in Hydrofluosilicic Acid, Deionized H_2O Rinse
B. Etched in Hydrofluosilicic Acid, Tap H_2O Rinse

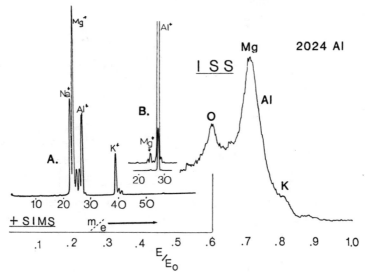

Figure 4.   ISS/SIMS Data From 2024 Aluminum Alloy, Degreased and Gentle Alkaline Clean. Inset shows SIMS Spectrum from Bulk Alloy.

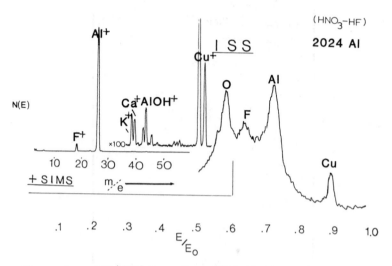

Figure 5.   ISS/SIMS Data From 2024 Aluminum Alloy, Etch in
HF/HNO$_3$.

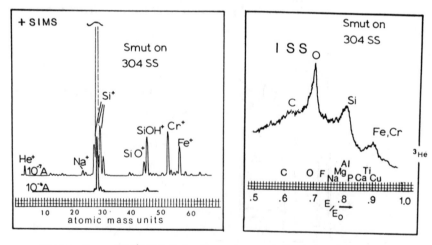

Figure 6.   ISS/SIMS Data From Smutted Stainless Steel

and is found to be mostly silicon and oxygen. Even when the surface
is visibly desmutted, often traces remain which are measurable by
ISS/SIMS. Smutted and de-smutted surfaces were examined by several
analytical techniques. The results of which are summarized in Table
IV.

Table IV
Species found on 304 stainless steel by surface analysis techniques

Technique	Smutted in H_2SO_4	De-smutted in $H_2SO_4-CrO_4$
AES	Si, O, Cl, C, Cu, Fe, Cr, Ni	O, C, S, Fe, Cr, Ni
ISS	Si, O, C, Cu, Fe, Cr	O, Na, Fe, Cr
+SIMS	Si^+, Na^+, SiO^+, $SiOH^+$, Cu^+, CH_n, Fe^+, Cr^+, Ni^+, Cu^+	Na^+, K^+, Si^+, S^+, OH^+, CH_n^+, Fe^+, Cr^+, Ni^+
-SIMS	$C_nH_n^-$, O^-, OH^-, SiO^-, SiO_2^-, SiO_3^-, SiF^-, $SiOF^-$, SiO_2F^-, FeO_2^-	$C_nH_n^-$, O^-, OH^-, Cl^-, SiO^- (greatly reduced) CrO^-, CrO_2^-, CrO_3^-, FeO_2^-
XPS	C^a, O, S, Si^b, N, Cu, Fe, Cr, Ni	C, O, Fe, Cr, Ni

[a] More than one form.

[b] Oxide form.

Chemistry and Structure of Adhesives

SIMS is very sensitive to surface molecular structure, showing
fragmentation pattern changes even on the same material but given
different treatment. Figure 7 shows SIMS data for a commercial
two-part epoxy mixed under the same conditions and then divided into
two portions, one cured 24 hours at room temperature and the other
cured one hour at 250°F.
As can be seen, some larger fragments are seen in the sample
held at elevated temperature, and sodium has segregated to the
surface. Such segregation is very common in high temperature cured
specimens, where sodium is often found at the failure surface in an
adhesive failure mode. ISS/SIMS data from the adhesive side of a
titanium-epoxy failure interface from a tensile test specimen are
shown in Figure 8.
The fragmentation pattern is different (compared to the two
part epoxy) from this temperature sensitive tape epoxy and sodium is
seen at the failure interface. Sodium was also observed on the
matching titanium side of the specimen.

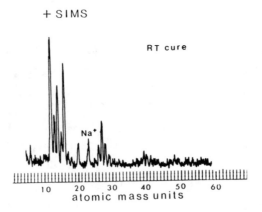

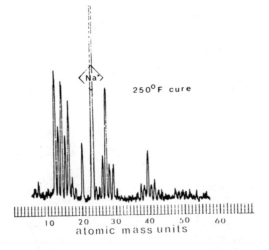

Figure 7.   SIMS Data From Two Part Epoxy Cured at Room
             Temperature for 24 Hours and at 250°F for One
             Hour.

Failure Surfaces

SIMS and the other complementary surface spectroscopies are extremely useful in determining just where an adhesive bonded structure actually separated. Following failure during service or test, it is not always obvious just where the failure took place. Often a failure is termed "adhesive" or interfacial just because the adherend appears to be "clean" (no adhesive). In actuality the failure may have occurred in a weak boundary layer very near an interface. A failure may also be initiated in one area and progress into another weaker area. Figure 9 shows a model of an adhesive bond and some of the features contributing to a fingerprint spectrum which pinpoints the exact locus of failure. Examples of these chemical "fingerprints" were shown earlier in ISS/SIMS spectra from alloy surfaces in which alloying and impurity element distributions were far different from bulk values. In addition, some of the interfacial regions show very characteristic spectra depending on past history or purposely added elements. For example, as mentioned earlier, bonds on metals or alloys which have been heated following etching or during processing often show weak boundary layer failures in which large quantities of alkali elements have migrated to the interface. Such a failure surface along with the original Ti-6Al-4V etched surface shown in Figure 10 originated with research on modeling of gold adhesion to titanium alloys. Following easy peel of the gold, the surface was found to produce a high sodium signal in the SIMS spectra which had not been observed in the original surface. The gold side of the failure also contained a large amount of sodium.

Purposely added elements also often help to pinpoint an exact locus of failure as illustrated in Figure 11. Here a failure is shown to have occurred near the primer-aluminum alloy interface as indicated by SIMS and other spectra which showed elements of the corrosion inhibitor (strontium chromate) on the failure surfaces. Similar work using SIMS has been used to infer environmental corrosion resistance[15] and to determine thickness of thin silanized surfaces[16]

SUMMARY

Ion beams provide useful information either as a diagnostic tool or as a precision etching method in adhesive bonding research. The combination of SIMS with complementary methods such as ISS or AES provides a powerful tool for elemental and limited structural characterization of metals, alloys and adhesives. The results shown here indicate that surface chemistry (and interface chemistry) can be decidedly different from bulk chemistry. Often it is this chemistry which governs the quality and durability of an adhesive bond. These same surface techniques also allow an analysis of the locus of failure of bonded materials which fail in service or test.

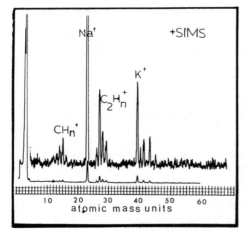 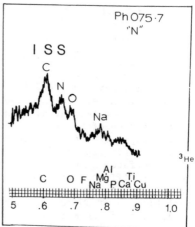

Figure 8.   ISS/SIMS Data From Tape Epoxy Adhesive Debonded
             From Titanium

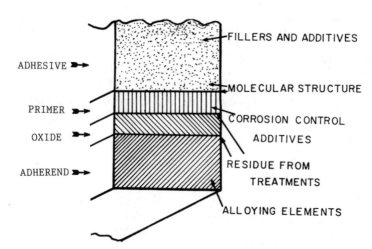

Figure 9.   Model of Adhesive Bond Showing Impurities and
             Additives

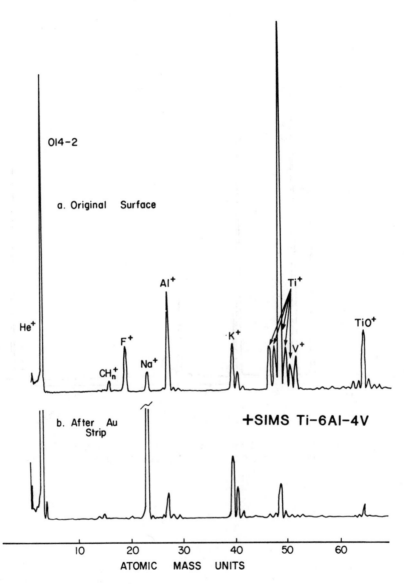

Figure 10.   SIMS Data from Ti-6Al-4V Alloy.
A.   Original Surface of Titanium Alloy
B.   Surface After Gold Stripped From Titanium Alloy

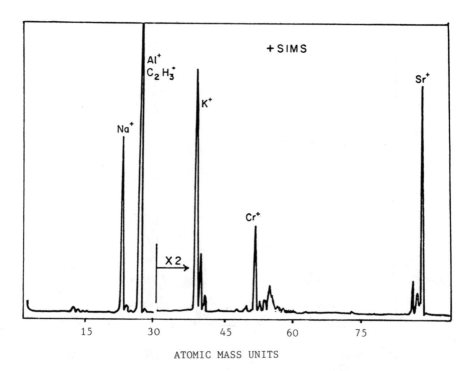

Figure 11.   SIMS Data From Aluminum Alloy Failure Surface
             Containing Corrosion Control Additive (Strontium
             Chromate)

Literature Cited

1. Baun, W. L.; <u>Appl. Surface Science</u>, 1980, 4, 291.
2. Powell, C. J.; <u>Appl. Surface Science</u>, 1978, 1, 143.
3. Benninghoven, A.; <u>Surface Science</u>, 1975, 53, 596.
4. McDevitt, N. T.; Baun, W. L.; Solomon, J. S.; <u>J. Electrochem</u>. <u>Soc</u>., 1976, 123, 1058.
5. McDevitt, N. T.; Baun, W. L.; Solomon, J. S.; AFML-TR-76-13, March 1976, Available NTIS.
6. McDevitt, N. T.; Baun, W. L.; Solomon, J. S. AFML-TR-75-122, October 1975, Available NTIS.
7. Sun, T. S.; Chen, J. M.; Venables, J. D.; Hopping, R.; <u>Appl. Surface Science</u>, 1978, 1 202.
8. Baun, W. L.; McDevitt, N. T.; Solomon, J. S. In: "Surface Analysis Methods for Metallurgical Applications"; ASTM STP 596, ASTM, Philadelphia, PA,, 1976, p. 86.
9. Baun, W. L.; AFML-TR-76-29, March 1976, Pt. I, Available NTIS
10. Baun, W. L.; McDevitt, N. T.; AFML-TR-76-29, May 1976, Pt. II, Available NTIS.
11. Baun, W. L.; McDevitt, N. T.; Solomon, J. S.; AFML-TR-76-29, October 1976, Pt. III, Available NTIS.
12. Baun, W. L.; McDevitt, N. T.; <u>J. Vac. Science Technology</u>, 1984, 2(2), 787.
13. Dyer, C. K.; Leach, J. S. L.; <u>J. Electrochem. Soc</u>., 1978, 125, 1032.
14. Baun, W. L.; <u>Surface Technology</u>, 11, 385. 1980.
15. Gettings, M.; <u>Kinloch, A. J.; J. Material Science</u>, 1977, 12, 2511.
16. Ross, M. R.; Evans, J. F.; In: "Proceedings of 7th Midland Macromolecular Symposium"; Leyden, D., Ed.; Gordon and Breach, 1980, pp. 99-123.

RECEIVED June 4, 1985

Author Index

Subject Index

Production by Hilary Kanter
Indexing by Deborah H. Steiner
Jacket design by Pamela Lewis

Elements typeset by Hot Type Ltd., Washington, D.C.
Printed and bound by Maple Press Co., York, Pa.